Intelligent Systems Reference Library

Volume 205

The aim of this series is to publish a Reference Library, including novel advances and developments in all aspects of Intelligent Systems in an easily accessible and well structured form. The series includes reference works, handbooks, compendia, textbooks, well-structured monographs, dictionaries, and encyclopedias. It contains well integrated knowledge and current information in the field of Intelligent Systems. The series covers the theory, applications, and design methods of Intelligent Systems. Virtually all disciplines such as engineering, computer science, avionics, business, e-commerce, environment, healthcare, physics and life science are included. The list of topics spans all the areas of modern intelligent systems such as: Ambient intelligence, Computational intelligence, Social intelligence, Computational neuroscience, Artificial life, Virtual society, Cognitive systems, DNA and immunity-based systems, e-Learning and teaching, Human-centred computing and Machine ethics, Intelligent control, Intelligent data analysis, Knowledge-based paradigms, Knowledge management, Intelligent agents, Intelligent decision making, Intelligent network security, Interactive entertainment, Learning paradigms, Recommender systems, Robotics and Mechatronics including human-machine teaming, Self-organizing and adaptive systems, Soft computing including Neural systems, Fuzzy systems, Evolutionary computing and the Fusion of these paradigms, Perception and Vision, Web intelligence and Multimedia.

Indexed by SCOPUS, DBLP, zbMATH, SCImago.

All books published in the series are submitted for consideration in Web of Science.

More information about this series at https://link.springer.com/bookseries/8578

Dalia Kriksciuniene · Virgilijus Sakalauskas
Editors

Intelligent Systems for Sustainable Person-Centered Healthcare

 Springer

Editors
Dalia Kriksciuniene
Vilnius University
Vilnius, Lithuania

Virgilijus Sakalauskas
Vilnius University
Vilnius, Lithuania

European Cooperation in Science and Technology
COST (European Cooperation in Science and Technology)

ISSN 1868-4394 ISSN 1868-4408 (electronic)
Intelligent Systems Reference Library
ISBN 978-3-030-79355-5 ISBN 978-3-030-79353-1 (eBook)
https://doi.org/10.1007/978-3-030-79353-1

This Springer imprint is published by the registered company Springer Nature Switzerland AG
The registered company address is: Gewerbestrasse 11, 6330 Cham, Switzerland

Preface

The collective monograph "Intelligent Systems for Sustainable Person-Centered Healthcare" establishes a dialog among the medical and intelligent system domains for igniting transition toward a sustainable and cost-effective health care, driven by advanced technologies. The Person-Centered Care (PCC) conceptual background of healthcare positions a person in the center of a healthcare system, instead of defining a patient as a set of diagnoses and treatment episodes. The PCC-based conceptual background triggers enhanced application of Artificial Intelligence (AI), as it dissolves the limits of processing traditional medical data records. The ambition of taking care of a person health by knowing life conditions, values, and expectations for nurturing own health adds new dimensions for making PCC operational.

The book discusses ability of intelligent healthcare system to monitor person health and improve quality of life. The monograph consists of three parts.

Part I discusses conceptual background of healthcare system, identifying major differences in required knowledge and linking to its sources while applying patient- or person- oriented frameworks of care. The extensive analysis and conceptualization of healthcare systems characterizes context for implementing AI approach for research and application of AI technologies.

Chapter 1 analyzes theoretical backgrounds of health care, it positions the person-centered care among the theoretical concepts of health care, shaping lifetime relationships among people and the medical institutions. The PCC concept suggests technological innovations and changes in providing healthcare services due to new requirements for inter-professional collaboration, application of scenario-based simulations and gaming for training of professionals at the healthcare institutions and redesigning their processes.

Chapter 2 defines the operational considerations of PCC and introduces the specific routines for its implementation based of the analysis of person narratives, negotiating and building healthcare plans, and linking to relevant documents influencing and characterizing person health.

Chapter 3 provides design and evaluation considerations for Person-Centred Care Implementation. It covers organizational process design for its implementation, applying intelligent technologies for processing information emanating from the

PCC operational routines. The measures, indicators, and methods for valuing health care and its effects are researched.

Chapter 4 discusses the customization for PCC intervention in a healthcare domain of pharmacy. It introduces the concept of PCC within pharmaceutical care delivery, explores its role as a part of multidisciplinary health services delivery teams. The role of health literacy, e-pharmacy service, and tele-pharmacy for implementing PCC is explored.

Part II provides research of efficiency evaluation, decision-making, and sustainability in person-centered health care.

Chapter 5 researches concept and models of shared decision-making as a framework of PCC. It focuses on multi-criteria decision-making techniques in healthcare settings. The ethical and practical considerations of shared decision-making in PCC and sensitive data emanating from patient narratives creates specific conditions and barriers for its implementation technologies.

Chapter 6 analyzes concepts, techniques, and methods for efficiency evaluation for health care. It provides considerations for evaluating the performance of healthcare system at European level followed by the enhancement of fundamental principles of methods of evaluation for healthcare institutions influenced by PCC.

Chapter 7 focuses on measurement methods and practices in health care applied to PCC interventions, considering the need and importance of measurement systems (outcomes and costs) to support and evaluate innovative health service delivery models.

Chapter 8 provides research of impact of human resources to the efficiency of PCC-based healthcare system and reveals different reactions, both resistance and support to the PCC-implied changes originating from the differences in institutional logics. The empirical research study in Ukraine and Poland reveals significance of motivators, education, and training for roles fulfilment of professionals.

Part III discusses intelligent systems and their application in health care. The four chapters provide research of AI application practice starting from adapting general data mining process to case studies of analytics and process design.

Chapter 9 discusses the performance of artificial intelligence methods and analysis of their application potential for different types of data sources of health care. The characteristics of methods, applied in the areas of data mining are revealed.

Chapter 10 concerns the problem of relevance of available information sources and substantiation of selecting efficient methods of AI for different problem areas. The chapter provides essential characteristics of methods and illustrates the aspects of their performance by experimental computations for the clinical data.

Chapter 11 researches application of mathematical methods for incorporating the information of a person and building mathematical models of decision support. The probabilistic methods, Bayesian networks are discussed for multidimensional decision support framework in medical domain.

Chapter 12 analyzes the intelligent process design. The case of hospital information management process illustrates the proposed methodology, where four types of knowledge-based UML dynamic models are generated by the transformation algorithms from the enterprise model.

The three parts of the book aim to reveal the challenges and tasks on intelligent for efficient health care in the PCC settings. It is an international teamwork of authors from 14 countries, represented by researchers and practitioners of data science and IT, health informatics, economics, healthcare quality, pharmacy, public health, professionals of healthcare management and practice.

Kaunas, Lithuania

Dalia Kriksciuniene
Virgilijus Sakalauskas

Acknowledgements This publication is based upon work from COST Action "European Net-work for cost containment and improved quality of health care- CostCares" (CA15222), supported by COST (European Cooperation in Science and Technology)

COST (European Cooperation in Science and Technology) is a funding agency for research and innovation networks. Our Actions help connect research initiatives across Europe and enable scientists to grow their ideas by sharing them with their peers. This boosts their research, career and innovation.

https://www.cost.eu

Contents

Part I
Person-Centred Healthcare System: Concept and Technological Requirements

Chapter 1
Analysis and Conceptualization of Healthcare Systems and Training in the Context of Technological Innovation and Personalization

Brenda Bogaert, António Casa Nova, Serap Ejder Apay, Zeynep Karaman Özlü, Paulo Melo, Jean-Philippe Pierron, Vítor Raposo, and Patricia Sánchez-González

Abstract This chapter will analyse personalization within the context of technological innovation. It will first of all clarify the conceptual terms used in the debate, in particular patient, person-centered and people-centered care and their various uses and limitations. It will then focus on specific issues of personalization and technology

B. Bogaert (✉)
Healthcare Values Chair, University Lyon III, IRPHIL, Salle 403, 18 rue Chevreul, 69007 Lyon, France
e-mail: brenda.bogaert@univ-lyon3.fr

Laboratory S2HEP University Lyon 1, Villeurbanne, France

A. C. Nova
Health School of Polytechnic Institute of Portalegre, Campus Politécnico 10, 7300-555 Portalegre, Portugal
e-mail: casanova@ipportalegre.pt

S. E. Apay
Department of Midwifery, Atatürk University, Erzurum, Turkey

Z. K. Özlü
Department of Surgical Nursing, Anesthesiology Clinical Research Office, Atatürk University, Erzurum, Turkey

P. Melo
Centre for Business and Economics Research, INESC Coimbra, University of Coimbra, Av. Dr. Dias da Silva 165, 3004-512 Coimbra, Portugal
e-mail: pmelo@fe.uc.pt

J.-P. Pierron
LIR3S (UMR CNRS-uB 7366), Université de Bourgogne, Dijon, France
e-mail: Jean-Philippe.Pierron@u-bourgogne.fr

Healthcare Values Chair, IRPHIL, Salle 403, 18 rue Chevreul, 69007 Lyon, France

V. Raposo
Centre for Business and Economics Research (CeBER), Centre of Health Studies and Research, University of Coimbra, Av. Dr. Dias da Silva 165, 3004-512 Coimbra, Portugal
e-mail: vraposo@fe.uc.pt

© The Author(s) 2022
D. Kriksciuniene and V. Sakalauskas (eds.), *Intelligent Systems for Sustainable Person-Centered Healthcare*, Intelligent Systems Reference Library 205,
https://doi.org/10.1007/978-3-030-79353-1_1

in emerging areas, notably in interprofessional practices and in medical training. This will allow greater understanding of both the possibilities and emerging tensions in the integration of personalization and technological innovation in healthcare systems from the training stage to its integration in various professional cadres.

Keywords Personalization · Person-centered care · People-centered care · Medical training · Interprofessional collaboration

1.1 Part I: Conceptual Perspectives—Patient, Person, and People-Centered Care

Personalization is a broad term that implies that the delivery of a medical service will be adapted to the needs of an individual patient. Today the notion has been developed in a number of areas to highlight advances in healthcare, from the development of personalized medicine to changing ideas on the doctor-patient relationship. These conceptions are changing both how we see and value the healthcare act and the patient's and healthcare provider's roles within it. Because of this, value remains an important term used throughout this chapter. While in health economics, value has often been understood in terms of cost, relational values are also advocated for within models such as person-centered care [1]. Embracing this plurality allows healthcare values to be analysed from both an economic and ethical perspective, congruently rather than separately, and it is within this framework that we will discuss value throughout this chapter.

Now that we have clarified our use of the word value, we will proceed with a discussion on personalization within the context of the doctor-patient relationship and in how we conceive of and organize healthcare. We will notably discuss the conceptual differences among patient, person-centered, and people-centered care and the challenges of implementing these different frameworks in the context of technological innovation, before moving onto to other aspects such as medical training and interprofessional collaboration.

1.1.1 Person-Centered Care

What does the term person (and not just patient) imply for healthcare systems? How does it change relationship and services, as well as healthcare evaluations? Can

P. Sánchez-González
Biomedical Engineering and Telemedicine Centre, Center for Biomedical Technology, ETSI Telecomunicación, Universidad Politécnica de Madrid, Madrid, Spain
e-mail: p.sanchez@upm.es

Centro de Investigación Biomédica en Red de Bioingeniería, Biomateriales y Nanomedicina (CIBER-BBN), Madrid, Spain taking into account the value of a person be a way to equilibrate the just and equitable relation between all the partners of a health and care system in the patient's interest? This section will first analyze the conceptual similarities and differences between person and patient-centered care, as these terms continue to be used concurrently and with some ambiguity across a wide variety of disciplines and contexts. We will then discuss thick and thin definitions of the person and how they may affect the conceptions of person-centered care. Finally, we will highlight some ongoing and emerging tensions in various ways that person-centered care has been conceptualized to help navigate its passage between fields as well as to avoid potential pitfalls. Within this discussion, we will also highlight specific issues relating to person-centeredness and emerging technology, specifically personalized medicine and e-medicine tools such as wearable medical devices, as the increased use of these technologies by patients and their healthcare providers present new conceptual and implementation challenges.

For 50 years, contestation of the paternalist model of medicine found support in the form of patient-centered care. The original term can be traced to Balint [2], who sought to encourage a mutual investment between the patient and healthcare providers. He sought to establish what he called an "overall diagnosis" which included everything the doctor knew and understood about the patient. In order to accomplish this, for Balint it was primordial for the patient to be understood as a unique human being. The development of the patient-centered care approach since then has helped take a holistic view of the person and establish a healthcare alliance between the patient and their healthcare provider. Terms such as relationship focused care, client-centered care, user-centered care, and recently person-centered care, have now been introduced. What is the advantage of changing the terminology from patient-centered (or another term) to person-centered care?

Changing from patient to person-centered care has several advantages. First of all, the patient-centered model brought new problems by encouraging a blind spot toward the relations and importance of care givers (from the physicians to nurses, from family to voluntary and patient associations). Person-centered care instead shows that relational values have been missing from current healthcare evaluations. These aspects are incompletely developed in most conceptions of patient-centered care, which effectively center the attention on the patient. Therefore, it seems necessary to build an inflection with person centered-care.

Secondly, while initially promising, the implementation of the patient-centered care model has not been wholly successful as it has not radically transformed the focus of the consultation, where the clinical gaze is based on the molecular and cellular basis of disease rather than the patient as a person 3. The reformulation of patient to person-centered care can therefore be seen as an ethical and public policy strategy. First of all, the important word change from "patient" to "person" may be able to focus the clinician's attention from the patient with a disease (understood as a pathology) to the person with an illness (the experience of being unhealthy for an individual). It may also help move care practice toward a respect for and acknowledgement of individuals in the context of their social lives and relationships with others. Although this has already been advocated by in patient-centered care, the word person necessarily widens the scope of the healthcare plan and the people involved. In addition, as the biomedical focus has been shown to be too narrow to encompass the variety of factors (family influences, environment, patient preferences, etc.) affecting healthcare outcomes, a new concept may enable healthcare institutions, providers, and researchers to rethink how to organize, promote, and evaluate care.

1.1.2 Different Ways to Conceptualize the Person: From a Thin Concept to a Thick Definition

At this point it will be necessary to specify what "person" means within the approach of person-centered care. It is possible to identify different criteria of the person which go from less to more, from the poorest to the richest, from the most quantitative to the more qualitative, from the most generalizable to the most irreplaceable. This analysis will make it possible to reveal competing uses of the individual in medicine (starting with so-called "personalized" medicine) and question their possible articulations.

(A)   The person may be identified to the physical and objective dimension of the human body and seems to assimilate the person and individual. A person may be the "the smallest denominator" that we can isolate: yesterday it was blood; now the gene and the genome. This physical signature is precise but also really poor in a subjective way, because the genome is more collective (biological relatives) more than a person. Personalized medicine can be found on this level.

(B)   The person can also mean "personality" in a psychological or psychiatric approach. This is an objective way if one wants to measure the qualitative of the subject in quantifiable data, since psychiatry follows somatic medicine. Thus the personality is objectively defined in universal classifications of diseases such as the DSM.

(C)   The person can be defined in the legal sense of the term, as a subject of law, capable of imputation of his acts, according to the ancient distinction inherited from Roman law between things (*res*) and persons (*persona*). This concept of persona makes it possible to establish a contractual dimension in the relationship of care, a contract between two subjects, or a subject and an institution. It also makes it possible to socialize the idea of illness and permits institutions to establish a cadre to protect the patient's rights.

(D)   The concept of person can be given even greater substance by recognizing persons an ends in themselves. "Things have a price, but only people have dignity," Kant will say, as we shall see below. This concept opens up the ethical scope of the person, which is protected by medical ethics and bioethics.

(E)   Finally, at the deepest or most consistent level, a final definition of the person makes him or her nucleus of a subjectivity, that of an irreplaceable self. The subject becomes the subject of a life engaged in all the dimensions of what makes a human existence, i.e. also in engaged in relations with others. It retraces the four dimensions of the person previously identified (genome, psyche, legal and moral status) which it personalizes and unfolds in the perspective of the subject and the aim for his or her life. The person here is not an individual (in-divisible) since he or she is not conceivable without his or her relationships with others. From this perspective, illness becomes an issue of existence with a relational scope: the biological fact of illness (disease) resounds like a biographical event.

1.1.3 Specific Issues for the Conceptualization of Person-Centered Care

Having elaborated the thin to thick definitions of the "person," we will now proceed by discussing some conceptualization possibilities and challenges in the person-centered care model.

(a) **Kantian perspectives of rights and duties in person-centered care**

The Kantian principle that we must accept each person as an end, and not as a means, underpins most concepts of person-centered care (level D highlighted above). This idea implies that persons are sources of agency (they have the capacity to capacity to act) and have dignity (they are ends) that must be respected. In healthcare this translates into giving a certain decision making power to the patient, thus promoting greater patient autonomy.

However, the implications of the Kantian principle also bring some important challenges to person-centered care, notably because seeing the person as an end also brings some implicit assumptions about both the person's rights and duties (the contract model highlighted in level C above). Person-centered models are based upon the premise that the person has a "right" to participate in healthcare decision making; however, if patients participate, there is also an implicit assumption that the patient will adhere to the treatment plan agreed in the healthcare alliance (thus implying a "duty" or a "responsibility" to adhere). Indeed, advocacy for person-centered models center on the possibility that they can be more cost-effective as patients will be more likely to adhere to a treatment plan in which they were actively involved. This means that we explicitly invoke patient rights to participate and implicitly patient duties to adhere. It is not clear which priority should be given the most attention in person-centered models, because the discussion centers both on respecting the patient's choices and reducing costs. The risk of tying together rights and duties is not only a lack of conceptual clarity. If we do not openly discuss these tensions and agree on what exactly we are asking of patients, we could risk ending up with another paternalistic model repackaged as person-centered.

(b) **Paul Ricoeur and Martha's Nussbaum's ideas on capabilities and vulnerabilities**

Some concepts by Paul Ricoeur 4 and Martha Nussbaum [5] have been integrated in concepts of person-centered care (level E highlighted above) by recognizing that patients are both capable and vulnerable. The advantage of these conceptions is to move beyond advocacy for individual patient autonomy and to both recognize the person as an end (the Kantian principle) and that people are in need of a facilitating environment due to their specific vulnerabilities as patients. As level E highlighted, it recognizes that the person is a subject of life but is also engaged in their relationships with others. Nussbaum's version of the capability approach for instance defends the idea that individual people should decide for themselves what they wish to be and to do (their capabilities), but she also recognizes that we need others (a facilitating environment) to develop and put into action our life projects. In models such as Entwistle and Watt's person-al capabilities approach 6, healthcare will therefore be organized to not only respect the individual needs, values, or priorities, but it will also encourage the healthcare provider to help cultivate the person's capabilities.

Likewise, Paul Ricoeur's ethical approach defines the person as both capable and vulnerable by showing how our identities are articulated in relationships and meditated via institutions. A central idea of Ricoeur's philosophy is the importance of our narrative identity, which helps us to create cohesion in our lives. Ricoeur's theory of narrative identity has inspired Charon [7] narrative medicine approach as well as the Gothenburg person-centered care model [8]. Applying Ricoeur's intuitions to a person-centered perspective encourages healthcare providers to pay attention to and document patient narration so that they can work with them in the context of their overall lives and to identify what is important to them. In addition, by recognizing persons in the healthcare alliance as both vulnerable and capable, it also signals the interrelationship and interdependency between healthcare providers, patients, and their families.

This approach may be time-consuming and/or costly in at least some temporalities of healthcare organization, as it involves considerable investment in working with patients to cultivate their capabilities and/or to in the use of narrative-based approaches. It also remains difficult to advocate for in the face of realities such as increasing economic pressures on hospitals. However, Ricoeur reminds us that ethics and economics are symbolic mediations of our institutions and must be debated congruently rather than separately to enable creative institutional change [9]. Already quantifying costs and benefits for person-centered care has shown some promising results, both in terms of patient satisfaction [10] as well as reducing overall healthcare costs [11]. However in order to fully realize these ideals, healthcare organizations will need to rethink how healthcare acts can be measured and evaluated [1]. For future research, it will also be necessary to evaluate how costly it may be when we do not take care of the person, patient or caregiver.

1.1.4 Specific Implementation Challenges

Having highlighted the conceptual challenges inspiring person-centered care, this section will proceed by discussing several implementation challenges, in particular in relation to technology.

(a) *Integrating medical innovation and increasing complexity into person-centered care models*

Personalized medicine remains an example of the complexity of integrating medical innovation into the person-centered healthcare model. Personalized medicine (also known as stratified medicine) specifically targets and adapts a treatment based on individual characteristics, in particular genomic factors. It is not a question of creating individual medication or strategies for each patient, but rather to establish subgroups that will allow treatment adaptations based upon subgroups of patient profiles [12]. Technology plays an important role in the realization of personalized medicine through the so-called omics technologies, which may allow diagnosis of a disease at the molecular level and to then use that information to develop targeted treatments specifically for that specific patient [13]. As highlighted in level A above, this focus of the person is at its thinnest, as it has shifted attention to the genomic level.

In order for personalized medicine to integrate the qualities of person-centered models, it will therefore be necessary to widen the perspective to the overall person and to pay greater attention to how individual behaviors may affect treatment efficacy. Personalized medicine will also need to resolve how patients can participate in healthcare decision making in face of increasing specialization and the integration of large amounts of data, as the technical complexity of personalized medicine is already disrupting the practice of medicine by bringing new challenges for the healthcare provider, who is expected not only to master molecular biology but also to have working knowledge of bioinformatics and biostatistics [14]. In this situation, it is unclear how patients will be able to participate in their treatment decisions other than as "sources" of information. Technology may also place some patients in a situation of greater vulnerability and/or dependence and prohibit or discourage them from participating. The risk therefore of integrating personalized medicine in person-centered care is to ignore—or at least minimize—the holistic perspective of the patient, as well as how technology may introduce new vulnerabilities in the healthcare alliance.

(b)   *Healing fractured healthcare infrastructures*

An important issue going forward in person-centered models remains of how to better coordinate and organize care among different specialties. George Engel's biopsychosocial model [15] inspired and provided the methodology for many patient-centered care models to take a holistic view of the patient. The results however have encouraged a certain dichotomy in care organization, such as regulating the biological aspect to the doctor, the psychological aspect to the psychologist, or the social aspect to social workers (such as in level B above) and have thus created a division of labor inside hospitals. Care organization has become fractured among different professionals in the patient's journey [16]. Person-centered models will need to resolve this implementation difficulty in fractured healthcare systems, such as with the designation of a reference person, to better accompany the patient in their healthcare journey. Formulations of person-centered care will also need to pay attention to how technology—from wearable medical devices, shared medical records, or the patient use of digital spaces—have affected care. While the use of technology represents an opportunity for the person-centered perspective to take into account the patient's, healthcare provider's and family's digital literacy, preferences, and vulnerabilities, the question still remains of how it can integrate these attentions across health systems and specialties.

(c)   *Political issues related to person-centered care*

A final issue will need to be highlighted before continuing our discussion onward to formulations beyond person-centered care. As discussed by Kreindler [17], the language and conceptualization of patient and person centered healthcare is not neutral. It has been notably been used to gain negotiating room or to reaffirm political positions. For instance, healthcare managers tend to emphasize the service/system level of person-centered care, using it as a pressure tool to influence employee behavior. On the other hand, professional groups have used PCC language to claim that their practices are patient-centered while others are not, thereby perpetuating ongoing political debates on hospital hierarchies (such as between doctors and nurses). Thirdly, while patient groups use PCC language to advocate for inclusion in healthcare decision making, they also may also use it to further their own interests and influence. It will be important to be vigilant of these political issues in formulations of person-centered care to avoid their deformation by political groups, but also to guard their ethical core (care *with* and *for* the patient). As suggested by Kreindler, this can be done by guiding the conversation and healthcare organization toward shared interests as well as valuing the epistemic contribution of each group in the design of person-centered healthcare programming. We will return to this issue when we discuss interprofessional collaboration in a later part of the chapter. For now, let us move toward an emerging concept being discussed in this debate, people-centered care, to understand what it might bring to this discussion.

1.1.5   People-Centered Care

This section will clarify what a move from person-centered to people-centered care implies for health systems. It will also discuss its implications in terms of healthcare innovation and cost-effectiveness. To start with, putting people at the centre of health services is a core aspect of health systems. It implies that services are organised around people's needs and expectations to make them more socially relevant and responsive whilst also producing better results [18, 19]. Good governance places people, rather than care providers, at the center of health systems [20, 21]. One of the core principles of good health governance is responsiveness so that institutions and processes can serve all stakeholders [22, 23] but also, as one of the three goals of the health system, to meet people's legitimate non-health expectations about how the system treats them [24].

People-centred care focuses on health needs, enduring personal relationships, comprehensive, continuous and person-centred care, responsibility for the health of all in the community along the life cycle and responsibility for tackling determinants of ill-health. In this model, people are partners in managing their health and that of their community [19]. According to the WHO [19, 25], people-centred care is focused and organised around people and their needs, rather than around diseases. Therefore, disease prevention and management are seen as necessary but insufficient to address people and communities' needs and expectations [26].

People-centred care is defined as an approach to care that intentionally adopts different stakeholders (individuals, healthcare providers, families and communities) perspectives as participants *in* and beneficiaries *of* trusted health systems that respond to their needs and preferences [26, 27]. People-centred care requires *people* empowerment, through education and support, to help citizens take more responsibility and participate in their care. In this conception, people should act as partners both in managing their health *and* in their community [19]. This approach may benefit individuals and their families, health professionals, communities, and health systems [26].

The main advantage of a conception built on people-centred care is that health will be understood as more than just healthcare. It recognizes that there is a wide range of social determinants (physical environment, social and economic factors, health care, and health behaviours) that influence how long and how well we live [28, 29]. For instance, a study by the Institute for Clinical Systems Improvement [30] has pointed out that the healthcare dimension (access to care and quality of care) only accounts for 20% of population health, emphasizing the need for investments on other social determinants dimensions.

To provide the context for people-centred and integrated health services, the WHO [26] has developed a conceptual framework representing the relationships between the different parts of the health ecosystem (Fig. 1.1). The proposed framework acknowledges the importance of intersectoral action in tackling the structural determinants of health and the close collaboration required between different sectors (health, social care, education) and other local services.

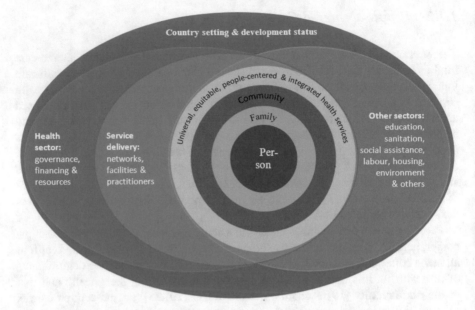

Fig. 1.1 WHO conceptual framework for people-centred care and integrated health services (Adapted from WHO 26)

From the individual and family perspective, the potential benefits of this concept include increased satisfaction with care and better relationships with care providers, improved access and timeliness of care, empowerment through improved health literacy and decision-making skills that promote independence, shared decision-making with professionals, increased involvement in care planning, the reinforcement of the ability to self-manage and control long-term health conditions, and better coordination between different care settings [26].

1.1.6   Innovation in Healthcare and People-Centered Care

Having conceptualized people-centered care and its advantages to help move forward by incorporating a community perspective, this section will focus on how citizens' needs and expectations be furthered through innovation. According to Santana et al. [31], people-centered healthcare systems need to be responsive to their specific contexts and identify priorities while encouraging innovation (PCC). For instance, the World Health Organization's (WHO) Health Innovation Group considers that health innovation can develop and deliver new or improved health policies, systems, products, technologies, services and delivery methods to improve people's health]32]. For its part, the Copenhagen Health Innovation [33] considers health innovation includes merging knowledge, development and technological opportunities with practice to improve the quality of life for patients and citizens. Finally, the Health Innovation Group [32], considers that health innovation comprises: (i) developing and implementing new or improved health policies, systems, products and technologies, as well as services and methods services that improve people's health, (ii) responding to unmet needs by employing new ways of thinking and working with a special focus on the needs of vulnerable populations, (iii) adding value in the form of improved efficiency, effectiveness, quality, safety and/or affordability, (iv) the ability to serve as preventive, promotive, therapeutic, rehabilitative and/or assistive care.

In a context of training and education, innovation can help match the needs of health and social care sectors with study programs aimed at: (i) the identification of needs and challenges in the health and social care sectors; (ii) the development of ideas for solutions and interdisciplinary innovation projects; (iii) testing ideas in close collaboration with practice (organisations, health services, institutions); (iv) the analysis of solutions, creating a basis for implementation decisions; and (v) implementation support [33].

Organisations can also make changes in their working methods, use of production factors and output types, improving productivity and performance using different types of innovation, including product, process, organisational and marketing [34]. Innovation can occur at various levels, with new or changed products or services, technical innovations can be technology-based (e.g., electronic medical records), process-based (e.g., care coordination), product-based (e.g., shared decision-making tools), or administrative (e.g., changing workflows, organisational structure, or human resource management) [35].

The vast majority of health systems in Europe have an essential public service component, namely through national health services. Public sector innovation in these contexts will involve creating, developing and implementing practical ideas to achieve a public benefit [36]. According to Paul and Per [37] this includes service innovation, service delivery innovation, administrative and organisational innovation, conceptual innovation, policy innovation, and systemic innovation.

Mulgan [36] argues that in the public sector, innovation can translate into new ways of managing organisations (such as public-private partnerships), new practices of rewarding people (such as performance-related pay), new ways of communicating (for instance, through ministerial blogs), policy or service innovations and innovations in other fields (e.g. e-voting) and international affairs (e.g. prepayments for new vaccines). According to him, some innovations warrant systemic change, such as creating a national health service or the move to a low carbon economy.

Concurrently, Hernandez et al. [38] have proposed a PCC framework defined as (1) effective leadership, (2) internal and external motivation to change, (3) clear and consistent organisational mission, (4) aligned organisational strategy, (5) robust organisational capability, and (6) continuous feedback and organisational learning. Several methodologies within this framework can be proposed, such as improving coordination and access to healthcare and services [39, 40] through a mixed methodology such as product/service, service delivery, process, administrative and organisational, conceptual, policy or systemic initiatives which can help integrate different stakeholders (health professionals, persons and communities, etc.). One promising methodology is also developing Health Labs to improve healthcare quality and facilitate cost containment [41].

Regardless of the implementation strategy, however, key questions that should be asked in a people-centered perspective include what is required support its implementation (including staff and infrastructure costs per person and at the aggregate level), what is the impact on service utilization (such as prevention of visits to urgent cure and unplanned stays), what is the impact on self-management and care prevention, and how improvements in physical health and medicine can be optimized [41].

While we could find no systematic review that focuses explicitly on PCC and innovation, the cost-effectiveness of person-centred health systems can evaluated via a value-based approach [42]. Examples of this include value-based healthcare, as introduced by Porter [43], value for money as defined by Smith [44] and Fleming [45] and economic evaluation [46]. All of these represent frameworks that can help evaluate costs and consequences for the questions raised relative to PCC and Health Labs. However, they also lead to new questions about how to feasibly evaluate people-centered care on the ground. One again it returns to us questions of value and how we can be responsive to people's needs and expectations.

1.2 Part II: Systemic Analysis of Personalization and Technology in the Context of Interprofessional Collaboration and Medical Training

Having clarified emerging conceptual perspectives of personalization, this section will seek to analyze the implications of personalization in different areas of healthcare practice. It will start with an analysis of the issues surrounding integration of personalization and technology in the context of medical training. Using the example of surgical training, we will show the concrete possibilities of personalization and technological innovation when integrated into training. We will then focus on interprofessional collaboration, seeking to show how personalization can be integrated into healthcare system design and training, enabling professionals to better work together with and for their patients.

1.2.1 Personalization, Technology, and Medical Training

Technology allows personalization in several different medical areas, including in education, promotion of healthy lifestyles, rehabilitation therapies planning, and surgical interventions planning and performance. First of all, technology helps inform citizens about healthy habits and treatments. Person-centered care and health promotion are intertwined [41] and a growing evidence base suggests that integrating these approaches can improve health outcomes [47], whilst maintaining health care quality without increasing costs [48, 49]. To this end, several platforms have risen to allow and facilitate digital health promotion. These platforms focus on customization thanks to the incorporation of intelligent recommendation systems [50]. Technology, such as big data, wearables, or 3D printers, also enables the advancement of clinical techniques and research.

New technologies also have the potential to improve training of healthcare professionals. Previous research has proposed a definition of patient-centered medical education that is centered on patients, with patients and for patients, to ensure current and future doctors remain sensitive to all of the needs of the people they care for [51]. Education of medical professionals is now a key challenge in European Healthcare systems. Current pedagogical needs in medical education are closely related with (1) ethical concerns on learning and training in real patients and working with animals and (2) reconciling time devoted to learning with clinical practice, taking into account the European Work Time Directives.

The incorporation of technology into medical training can notably integrate these concerns, in particularly to build the capacity of surgical planners with minimally invasive techniques. They allow the clinician to make the best decision for each individual patient and to incorporate visualization techniques showing the original images, structures of interest, and/or surgical tools useful for the clinician. Their use in soft tissue surgeries is not yet fully extended (it is still a great challenge), but in trauma, dental and intraoperative radiotherapy interventions, we can find different solutions in clinical routines in hospitals and healthcare centers.

Technology-enhanced learning (TEL) also plays an important role in the transformation of medical learning processes, in particular by improving future healthcare professionals competencies through simulation. Whilst its focus has mainly been simulation for technical skills [52], cognitive skills are also among the key competences required for surgeons [53]. Simulation in medical education is the preferred route to address both pedagogical needs, and the learning curve can be shortened by learning outside of daily clinical practice. For instance, training on simulators has correlated with improved operative times and a greater efficiency of movement for different techniques.

Moreover, and without question, medical simulation through technology has the potential to replace the use of animals as human surrogates in medical training and to personalize as much as possible the pedagogical path to the needs of trainers. Examples such as the project MIS-SIM (Minimally Invasive Surgery Simulator Scenario Editor) empower teachers to create their own training scenarios rather than be constrained to a predefined set of tasks, allowing them to adapt to the needs of their students. As a simulator allowing users to create and share with the community virtual tasks personalized to the training needs of learners, it also allows users to engage with virtual reality based learning tools whilst remaining in complete control of the learning process.

1.2.2 Models of Collaborative Practice: The Doctor, Nurse, Midwife, and Patient

Having discussed the opportunities for personalization and technological integration in the context of medical (and in particular surgical) training, it will now be important to move into the field in order to understand how personalization can be facilitated through collaborative practices. As highlighted at the beginning of the chapter, in order for person or people-centered care to be achievable, it must integrate the shared contributions and expertise of different healthcare professionals. As highlighted by the Independent Nurse-Midwifery Practice, collaboration can be understood as "a process whereby healthcare professionals commonly manage the care" [54]. In other words, it includes the interaction of at least two professionals or disciplines organized in a common effort to solve or discover common problems with as much as patient participation as possible [55].

Practices based on interprofessional collaboration have the potential to reduce the cost of healthcare services while also improving patient outcomes and patient experience [56]. Effective communication and collaboration enables quality of care, notably by contributing to patient safety, reducing the length of hospitalizations, and enabling healthcare and social services to work together [57–62]. It also helps to increase confidence and respect among healthcare professionals, reducing competition and conflict, and in so doing enables healthcare professionals to share their knowledge and skills [56, 57,63–66]. Interprofessional collaboration enables healthcare professionals to understand one another better and helps in constituting a respectful environment for team members [56, 57, 63, 65, 66]. It enables professionals in the healthcare team to be in their professional roles, helping them to take common decisions and share the responsibility of providing care [56, 67], in particular through accountability, coordination, communication, assertiveness, autonomy, mutual trust, and respect [68].

In order to be achievable on the ground, it will be necessary to think how to provide tools from the training stage [69]. A report by the Institute of Medicine has defined five core competencies necessary for healthcare professionals, including patient-centered care, interdisciplinary teamwork, evidence-based practice, increased improvement and quality in practices of care, and informatics [70]. Medical, nursing, and midwifery students should receive the necessary education on how they can collaboratively work with one another and with their patients. Unfortunately, in current education methods, healthcare professionals use a discipline-specific method rather than favor an interdisciplinary approach.

In the field, there are also various obstacles to facilitate effective collaboration, notably role ambiguity, confusion, irregular hierarchical relationships, education differences, gender and cultural differences. While systems and processes have been designed to simplify communication and teamwork, these practices are not necessarily intuitive and must be learned and applied by all team members. Therefore, hospitals and education institutions should integrate these models into the initial and ongoing education programs of nurses, midwifes, doctors, and other healthcare service providers to instill a practice that can be used on the ground [71] to change existing working cultures, especially in critical care [70].

Healthcare law from the United States in 2010 contributed to the renewal in the education of healthcare professionals and in the development of new interprofessional care models. The American Institute of Medicine notably recommended nurses to be leaders of the healthcare team working in cooperation with other healthcare professionals. Based upon these new conceptions, nursing education should therefore prioritize leadership, teamwork, and cooperation skills [72]. The University of Virginia Center of Academic Strategic Partnerships for Interprofessional Research and Education (ASPIRE) has developed a model to overcome difficulties in interprofessional education by focusing on practical tools, leadership, and relational factors [73]. Improvements in patient safety and quality of care were observed in practices in which this model was applied [73].

However, despite these promising developments, there still remains a gap between the goals and the reality of practices in higher education. For instance, while the Turkish Council of Higher Education (CoHE) states that by the end of their undergraduate education, students should have already developed these kind of collaboration skills, the curriculum is currently organized as a one-profession education. To mitigate this difficulty, students from different healthcare areas should come together more often and receive education together in order to acquire the required skills and professional standards needs for accreditation [74].

Another methodology which can help facilitate collaboration is simulation. As one of the most important determinants of the ability to transfer what students learn in the laboratory to the clinical environment, it can also reduce the reality shock they experience once in the field [75]. The most frequently used simulation methods include anatomical models, task trainers, role-play, games, computer assisted instruction (CAI), virtual reality, low-fidelity to high-fidelity mannequins, and standardized patients 76. Promising results have been observed in these simulation activities in developing team members' attitudes toward collaborative care [77].

To conclude this section, healthcare services require different professional groups to work together to increase the quality of care and patient satisfaction, in order to reduce costs as well as medical errors, and to increase employees' work satisfaction and efficacy. For this reason, developing collaboration-based practice skills remain a priority in healthcare training. The use of technologies such as virtual reality and computer assisted instruction can help facilitate this goal; however collaboration will also need to be prioritized across trainings and professions to facilitate its implementation on the ground.

1.3 Conclusion

Complexities such as surgical training and interprofessional collaboration have shown that technology and the political, economic, and ethical issues concerning personalization go hand in hand and must be dealt with congruently rather than separately. What this chapter has shown us is that healthcare implementation and training will need to be designed with and for all actors concerned, should it be the clinician, patient, family, or the community, in order to ensure that it responds to their needs and priorities. Only then can we realistically talk about the integration of technology and hope to advance toward personalized healthcare systems from a patient, person, and people-centered perspective.

Acknowledgements This publication is based upon work from COST Action "**European Network for cost containment and improved quality of health care-CostCares**" (CA15222), supported by COST (European Cooperation in Science and Technology)

COST (European Cooperation in Science and Technology) is a funding agency for research and innovation networks. Our Actions help connect research initiatives across Europe and enable scientists to grow their ideas by sharing them with their peers. This boosts their research, career and innovation.

https://www.cost.eu

Funded by the Horizon 2020 Framework Programme of the European Union

References

1. Pierron, J.-P., Vinot, D.: The meaning of value in "person-centred" approaches to healthcare. EJPCH **8**, 193 (2020). https://doi.org/10.5750/ejpch.v8i2.1842
2. Balint, E.: The possibilities of patient-centered medicine. J. R. Coll. Gen. Pract. **17**, 269–276 (1969)
3. Miles, A., Asbridge, J.: Person-Centered healthcare—moving from rhetoric to methods, through implementation to outcomes. EJPCH **5**, 1 (2017). https://doi.org/10.5750/ejpch.v5i1.1353
4. Ricœur, P., Blamey, K.: Oneself as Another. Nachdr. Univ. of Chicago Pr, Chicago, Ill (2008)
5. Nussbaum, M.C.: Creating capabilities the human development approach. Orient Blackswan, New Delhi (2011)
6. Entwistle, V.A., Watt, I.S.: Treating patients as persons: a capabilities approach to support delivery of person-centered care. Am. J. Bioeth. **13**, 29–39 (2013). https://doi.org/10.1080/15265161.2013.802060
7. Charon, R.: Narrative Medicine: Honoring the Stories of Illness, 1, paperback Oxford University Press, Oxford (2008)
8. Ekman, I., Swedberg, K., Taft, C., Lindseth, A., Norberg, A., Brink, E., Carlsson, J., Dahlin-Ivanoff, S., Johansson, I.-L., Kjellgren, K., Lidén, E., Öhlén, J., Olsson, L.-E., Rosén, H., Rydmark, M., Sunnerhagen, K.S.: Person-centered care—ready for prime time. Eur. J. Cardiovas. Nurs. **10**, 248–251 (2011). https://doi.org/10.1016/j.ejcnurse.2011.06.008. Engel, G.L.: From biomedical to biopsychosocial: being scientific in the human domain. Psychosomatics **38**, 521–528 (1997). https://doi.org/10.1016/S0033-3182(97)71396-3
9. Pierron, J.-P.: De l'agent économique à l'homme capable. Une critique de l'économisme à partir de l'herméneutique critique de Paul Ricœur. errs 9:124–137 (2019). https://doi.org/10.5195/errs.2018.429
10. Gyllensten, H., Koinberg, I., Carlström, E., Olsson, L.-E., Olofsson, E.H.: Economic analysis of a person-centered care intervention in head and neck oncology. Eur. J. Public Health **27** (2017). https://doi.org/10.1093/eurpub/ckx187.336

11. Pirhonen, L., Bolin, K., Olofsson, E.H., Fors, A., Ekman, I., Swedberg, K., Gyllensten, H.: Person-centred care in patients with acute coronary syndrome: cost-effectiveness analysis alongside a randomised controlled trial. PharmacoEconomics Open **3**, 495–504 (2019). https://doi.org/10.1007/s41669-019-0126-3
12. Giroux, E.: Les enjeux normatifs de la médecine personalisée. In: Les valeurs du soin. Enjeux éthiques, économiques et politiques, Paris, Seli Arslan (2018)
13. Nice, E.C.: Challenges for omics technologies in the implementation of personalized medicine. Expert Rev. Precis. Med. Drug Dev. **3**, 229–231 (2018). https://doi.org/10.1080/23808993.2018.1505429
14. Cox, S., Rousseau-Tsangaris, M., Abou-Zeid, N., Dalle, S., Leurent, P., Cutivet, A., Le, H.-H., Kotb, S., Bogaert, B., Gardette, R., Baran, Y., Holder, J.-M., Lerner, L., Blay, J.-Y., Cambrosio, A., Tredan, O., Denèfle, P.: La médecine de précision en oncologie : challenges, enjeux et nouveaux paradigmes. Bull. Cancer **106**, 97–104 (2019). https://doi.org/10.1016/j.bulcan.2019.01.007
15. Engel, G.L.: From biomedical to biopsychosocial: being scientific in the human domain. Psychosomatics **38**, 521–528 (1997). https://doi.org/10.1016/S0033-3182(97)71396-3
16. Weber, J.-C.: La consultation. Presses Universitaires de France, Paris (2017)
17. Kreindler, S.A.: The politics of patient-centred care. Health Exp. **18**, 1139–1150 (2015). https://doi.org/10.1111/hex.12087
18. World Health Organization: People at the centre of health care: harmonizing mind and body, people and systems. WHO Regional Office for the Western Pacific, Manila (2007)
19. World Health Organization: The World Health Report 2008—primary Health Care (Now More Than Ever) (2008)
20. World Health Organization: Governance for health equity: taking forward the equity values and goals of Health 2020 in the WHO European Region (1998a)
21. World Health Organization: Health for All in the 21st Century. World Health Organization (1998b)
22. United Nations Development Programme: Governance for Sustainable Human Development—A UNDP Policy Document. United Nations Development Programme UNDP policy document, New York (1997)
23. United Nations Development Programme: Reconceptualising governance. Management Development and Governance Division, Bureau for Policy and Programme Support, UNDP (1997b)
24. World Health Organization: The World Health Report 2000—Health systems: improving performance (2000)
25. World Health Organization: People-centred care in low and middle income countries (2010a)
26. World Health Organization: WHO global strategy on people-centred and integrated health services (2015b)
27. World Health Organization: People-centred and integrated health services: an overview of the evidence. World Health Organization (2015a)
28. Kindig, D., Stoddart, G.: What is population health? Am. J. Public Health **93**, 380–383 (2003). https://doi.org/10.2105/AJPH.93.3.380
29. Solar, O., Irwin, A.: A conceptual framework for action on the social determinants of health. WHO Document Production Services (2010)
30. Institute for Clinical Systems Improvement: Going Beyond Clinical Walls: Solving Complex Problems (2014)
31. Santana, M.J., Manalili, K., Jolley, R.J., Zelinsky, S., Quan, H., Lu, M.: How to practice person-centred care: a conceptual framework. Health Expect **21**, 429–440 (2018). https://doi.org/10.1111/hex.12640
32. World Health Organization: WHO Health Innovation Group (2021). http://www.who.int/life-course/about/who-health-innovation-group/en/. Accessed 11 Jan 2021
33. Copenhagen Health Innovation.: Health innovation thought education (2020). https://copenhagenhealthinnovation.dk/health-innovators/. Accessed 11 Jan 2021

34. OECD: Oslo Manual—Guidelines for Collecting and Interpreting Innovation Data, OECD Publishing (2005)
35. Crossan, M.M., Apaydin, M.: A multi-dimensional framework of organizational innovation: a systematic review of the literature: a framework of organizational innovation. J. Manage. Stud. **47**, 1154–1191 (2010). https://doi.org/10.1111/j.1467-6486.2009.00880.x
36. Mulgan, G.: Innovation in the Public Sector: How can organisation better create, improve, and adapt? (2014)
37. Paul, W., Per, K.: Innovation in public sector services: entrepreneurship, creativity and management. Edward Elgar, Cheltenham, Northampton, MA, UK (2008)
38. Hernandez, S.E., Conrad, D.A., Marcus-Smith, M.S., Reed, P., Watts, C.: Patient-centered innovation in health care organizations: a conceptual framework and case study application. Health Care Manage. Rev. **38**, 166–175 (2013). https://doi.org/10.1097/HMR.0b013e31825e 718a
39. Bhattacharyya, O., Blumenthal, D., Stoddard, R., Mansell, L., Mossman, K., Schneider, E.C.: Redesigning care: adapting new improvement methods to achieve person-centred care. BMJ Qual. Saf. **28**, 242–248 (2019). https://doi.org/10.1136/bmjqs-2018-008208
40. Sharma, T., Bamford, M., Dodman, D.: Person-centred care: an overview of reviews. Contemp. Nurse **51**, 107–120 (2015). https://doi.org/10.1080/10376178.2016.1150192
41. Lloyd, H.M., Ekman, I., Rogers, H.L., Raposo, V., Melo, P., Marinkovic, V.D., Buttigieg, S.C., Srulovici, E., Lewandowski, R.A., Britten, N.: Supporting innovative person-centred care in financially constrained environments: the WE CARE exploratory health laboratory evaluation strategy. IJERPH **17**, 3050 (2020). https://doi.org/10.3390/ijerph17093050
42. Paparella, G.: Person-centred care in Europe: a cross-country comparison of health system performance, strategies and structures (2016)
43. Porter, M.E.: What is value in health care? N. Engl. J. Med. **363**, 2477–2481 (2010). https://doi.org/10.1056/NEJMp1011024
44. Smith, P.: Measuring value for money in healthcare: concepts and tools (2009)
45. Fleming, F.: Evaluation methods for assessing Value for Money (2013)
46. Drummond, M.F., Sculpher, M.J., Claxton, K., Stoddart, G.L., Torrance, G.W.: Methods for the Economic Evaluation of Health Care Programmes. Oxford University Press (2015)
47. Hansson, E., Ekman, I., Swedberg, K., Wolf, A., Dudas, K., Ehlers, L., Olsson, L.-E.: Person-centred care for patients with chronic heart failure—a cost–utility analysis. Eur. J. Cardiovasc. Nurs. **15**, 276–284 (2016). https://doi.org/10.1177/1474515114567035
48. Fors, A., Ekman, I., Taft, C., Björkelund, C., Frid, K., Larsson, M.E., Thorn, J., Ulin, K., Wolf, A., Swedberg, K.: Person-centred care after acute coronary syndrome, from hospital to primary care—a randomised controlled trial. Int. J. Cardiol. **187**, 693–699 (2015). https://doi.org/10.1016/j.ijcard.2015.03.336
49. Fors, A., Swedberg, K., Ulin, K., Wolf, A., Ekman, I.: Effects of person-centred care after an event of acute coronary syndrome: two-year follow-up of a randomised controlled trial. Int. J. Cardiol. **249**, 42–47 (2017). https://doi.org/10.1016/j.ijcard.2017.08.069
50. Hors-Fraile, S., Rivera-Romero, O., Schneider, F., Fernandez-Luque, L., Luna-Perejon, F., Civit-Balcells, A., de Vries, H.: Analyzing recommender systems for health promotion using a multidisciplinary taxonomy: a scoping review. Int. J. Med. Informat. **114**, 143–155 (2018). https://doi.org/10.1016/j.ijmedinf.2017.12.018
51. Hearn, J., Dewji, M., Stocker, C., Simons, G.: Patient-centered medical education: a proposed definition. Med. Teach. **41**, 934–938 (2019). https://doi.org/10.1080/0142159X.2019.1597258
52. Schreuder, H.W.R., Oei, G., Maas, M., Borleffs, J.C.C., Schijven, M.P.: Implementation of simulation in surgical practice: minimally invasive surgery has taken the lead: the Dutch experience. Med. Teach. **33**, 105–115 (2011). https://doi.org/10.3109/0142159X.2011.550967
53. Madani, A., Vassiliou, M.C., Watanabe, Y., Al-Halabi, B., Al-Rowais, M.S., Deckelbaum, D.L., Fried, G.M., Feldman, L.S.: What are the principles that guide behaviors in the operating room?: Creating a framework to define and measure performance. Ann. Surg. **265**, 255–267 (2017). https://doi.org/10.1097/SLA.0000000000001962

54. Onan, A., Turan, S., Elçin, M.: Interprofessional training. Medical Trainer Handbook, pp 211–220
55. Selleck, C.S., Fifolt, M., Burkart, H., Frank, J.S., Curry, W.A., Hites, L.S.: Providing primary care using an interprofessional collaborative practice model: what clinicians have learned. J. Prof. Nurs. **33**, 410–416 (2017). https://doi.org/10.1016/j.profnurs.2016.11.004
56. Golom, F.D., Schreck, J.S.: The journey to interprofessional collaborative practice. Pediatr. Clin. North Am. **65**, 1–12 (2018). https://doi.org/10.1016/j.pcl.2017.08.017
57. World Health Organization: WHO Framework for action on interprofessional education and collaborative practice (2010b)
58. Brock, D., Abu-Rish, E., Chiu, C.-R., Hammer, D., Wilson, S., Vorvick, L., Blondon, K., Schaad, D., Liner, D., Zierler, B.: Republished: interprofessional education in team communication: working together to improve patient safety. Postgrad. Med. J. **89**, 642–651 (2013). https://doi.org/10.1136/postgradmedj-2012-000952rep
59. Karim, R., Ross, C.: Interprofessional education (IPE) and chiropractic. J. Can. Chiropr. Assoc. **52**, 76–78 (2008)
60. Raab, C.A., Will, S.E.B., Richards, S.L., O'Mara, E.: The effect of collaboration on obstetric patient safety in three academic facilities. J. Obstet. Gynecol. Neonatal. Nurs. **42**, 606–616 (2013). https://doi.org/10.1111/1552-6909.12234
61. Young, L., Baker, P., Waller, S., Hodgson, L., Moor, M.: Knowing your allies: medical education and interprofessional exposure. J. Interprof. Care **21**, 155–163 (2007). https://doi.org/10.1080/13561820601176915
62. Zanotti, R., Sartor, G., Canova, C.: Effectiveness of interprofessional education by on-field training for medical students, with a pre-post design. BMC Med. Edu. **15**, 121 (2015). https://doi.org/10.1186/s12909-015-0409-z
63. Al-Qahtani, M.F.: Measuring healthcare students' attitudes toward interprofessional education. J. Taibah Univ. Med. Sci. **11**, 579–585 (2016). https://doi.org/10.1016/j.jtumed.2016.09.003
64. Harden, R.M.: Interprofessional education: the magical mystery tour now less of a mystery: interprofessional education. Am. Assoc. Anatom. **8**, 291–295 (2015). https://doi.org/10.1002/ase.1552
65. Lawlis, T., Wicks, A., Jamieson, M., Haughey, A., Grealish, L.: Interprofessional education in practice: evaluation of a work integrated aged care program. Nurse Edu. Pract. **17**, 161–166 (2016). https://doi.org/10.1016/j.nepr.2015.11.010
66. Towle, A., Bainbridge, L., Godolphin, W., Katz, A., Kline, C., Lown, B., Madularu, I., Solomon, P., Thistlethwaite, J.: Active patient involvement in the education of health professionals: active patient involvement in education. Med. Edu. **44**, 64–74 (2010). https://doi.org/10.1111/j.1365-2923.2009.03530.x
67. Solomon, P., Salfi, J.: Evaluation of an interprofessional education communication skills initiative. Edu. Health (Abingdon) **24**, 616 (2011)
68. Kasperski, M.: Implementation strategies: 'Collaboration in primary care—family doctors and nurse practitioners delivering shared care.' (2000)
69. Zwarenstein, M., Reeves, S., Perrier, L.: Effectiveness of pre-licensure interprofessional education and post-licensure collaborative interventions. J. Interprof. Care **19**, 148–165 (2005). https://doi.org/10.1080/13561820500082800
70. Jeffries, P.R., McNelis, A.M., Wheeler, C.A.: Simulation as a vehicle for enhancing collaborative practice models. Crit. Care Nurs. Clin. North Am. **20**, 471–480 (2008). https://doi.org/10.1016/j.ccell.2008.08.005
71. Keleher, K.: Collaborative practice characteristics, barriers, benefits, and implications for midwifery. J. Nurse Midwifery **43**, 8–11 (1998). https://doi.org/10.1016/S0091-2182(97)00115-8
72. Dolce, M.C., Parker, J.L., Marshall, C., Riedy, C.A., Simon, L.E., Barrow, J., Ramos, C.R., DaSilva, J.D.: Expanding collaborative boundaries in nursing education and practice: the nurse practitioner-dentist model for primary care. J. Prof. Nurs. **33**, 405–409 (2017). https://doi.org/10.1016/j.profnurs.2017.04.002

73. Brashers, V., Haizlip, J., Owen, J.A.: The ASPIRE model: grounding the IPEC core competencies for interprofessional collaborative practice within a foundational framework. J. Interprof. Care **34**, 128–132 (2020). https://doi.org/10.1080/13561820.2019.1624513
74. Özata, K.: Determination of the Readiness of Students Studying in the Field of Health Sciences for Interprofessional Learning. (Master Thesis). Hacettepe University, Ankara (2018)
75. Bradley, P.: The history of simulation in medical education and possible future directions. Med. Edu. **40**, 254–262 (2006). https://doi.org/10.1111/j.1365-2929.2006.02394.x
76. Nehring, W.M., Lashley, F.R.: Nursing simulation: a review of the past 40 years. Simul. Gaming **40**, 528–552 (2009). https://doi.org/10.1177/1046878109332282
77. Yang, L.-Y., Yang, Y.-Y., Huang, C.-C., Liang, J.-F., Lee, F.-Y., Cheng ,H.-M., Huang, C.-C., Kao, S.-Y.: Simulation-based inter-professional education to improve attitudes towards collaborative practice: a prospective comparative pilot study in a Chinese medical centre. BMJ Open **7**, e015105 (2017). https://doi.org/10.1136/bmjopen-2016-015105

Chapter 2
Person-Centred Care, Theory, Operationalisation and Effects

Inger Ekman and Karl Swedberg

Abstract In healthcare systems patient engagement and care satisfaction are less than optimal. Different solutions have been proposed to recognise the patient in health care, including person-centred care. The University of Gothenburg Centre for Person-Centred Care (GPCC) steering committee formulated three 'simple routines' to initiate, integrate and safeguard person-centred care in daily clinical practice. These routines are: the patient narrative followed by an agreed health plan which is then safeguarded by documentation. Health care professionals need to know how health processes are strengthened in a relationship where patients are accepted as persons with their own will and emotions and in which individual responsibilities and capabilities are highlighted. A person-centred perspective uses ethics as a springboard. Such an ethical view can briefly be formulated by: "To aim for the good life, with and for others in just institutions". When the starting point is ethics and each person is understood as a unique individual, care actions will never be the same for each patient. By asking for the patients' understanding of the condition and treatment relative to their lives in general, professionals can understand what health, illness, treatment and care convey to patients and their relatives. The patient narratives are obviously very important in formulating the health plan. Controlled studies have found several benefits from implementing person-centred practices, including improved quality of life, maintained self-efficacy and reduced health costs.

Keywords Person-centred care · Patient narratives · Health plan

I. Ekman
Department of Health and Care Sciences, Sahlgrenska Academy University of Gothenburg, Gothenburg, Sweden
e-mail: inger.ekman@fhs.gu.se

K. Swedberg (✉)
Department of Molecular and Clinical Medicine, Sahlgrenska Academy, University of Gothenburg, Gothenburg, Sweden
e-mail: karl.swedberg@gu.se

© The Author(s) 2022
D. Kriksciuniene and V. Sakalauskas (eds.), *Intelligent Systems for Sustainable Person-Centered Healthcare*, Intelligent Systems Reference Library 205,
https://doi.org/10.1007/978-3-030-79353-1_2

2.1 Introduction

Healthcare systems need to be re-organized to provide high-quality care without increased costs to an ageing population with a high prevalence of chronic and long-term disorders [1, 2].

Currently, patient engagement and care satisfaction are less than optimal. Different solutions have been proposed to recognise the patient in health care, including person-centred care (PCC) initiatives.

Researchers and clinicians noted that PCC, emphasising patient-professional partnerships, has not been implemented in health care to a significant extent. Thus, in 2010, intending to test and implement PCC, an interdisciplinary group of clinical and non-clinical academics in Sweden created a research centre for the study of PCC in long-term illness: the University of Gothenburg Centre for Person-Centred Care (GPCC) [3, 4].

The GPCC steering committee formulated a position paper with three 'simple routines' to initiate, integrate and safeguard person-centred care in daily clinical practice [5].

The first routine serves to initiate a partnership by eliciting patient narratives, defined as the sick person's account of his or her perception of the illness and its impact on life.

In sharp contrast to medical narratives that reflect the process of diagnosing and treating the disease, the first routine captures the patients' suffering in the context of their daily lifeworld. The second routine implements the partnership principle using a commonly agreed personal health plan so that professionals, patients and relatives can work collaboratively to achieve the patients´ goals. The third routine safeguards the partnership by documenting the health plan accessible to both professionals and patients. This plan is often shown in the patient record. These three routines represent clinical tasks that professionals embark on and that patients and relatives perform in daily life. This PCC model is distinguished from other models by incorporating patient wishes and capabilities with care team support.

Such an approach is rooted in philosophical literature on [6–8].

2.2 The Patient—A Person

Health care professionals routinely understand and explain the patient from a medical perspective that focuses almost exclusively on biological and physiological factors. With this knowledge, the complex biology of human beings can be explained. However, even if this knowledge is important, health care professionals need to know more about how health processes are strengthened in a relationship in which patients are accepted as persons with their own will and emotions and in which individual responsibilities and capabilities are highlighted.

To understand another person, we must listen to what that person has to say. If the listening process in the communication is effective, the target person feels acknowledged and respected. Professionals must therefore listen to and understand patient needs and concerns. Such an approach implies that the professional must see patients as persons with a lifeworld (experiences and contacts that make up an individual's world), different but also similar to the professionals' in many ways.

Having such a perspective on understanding entails sensing, grasping or feeling the patient's experiences. In this context human vulnerability means being mindful and deeply affected by the suffering of others. This ability to have empathy may sometimes be painful to clinicians because a patient's suffering can be overwhelming, especially when there is insufficient time to meet each patient's needs. It could even mean attempts to abandon these feelings of vulnerability. However, vulnerability constitutes the notion of being human and postulates human capabilities. Human value is a concept of relationships: to understand and trust that one has a value one must be confirmed and recognised by another person.

The patient role can be part of an objective context as a closed system, i.e. the patient is reduced to only representing objective data of disease instead of an open system. In an open system the patient is a subject with autonomy and something important to contribute- the person is someone. Being a person and human being implies different roles, such as patient, teacher or beggar. Whatever position, the individual is treated as a person with dignity and respect. If only a diagnosis or objective data from tests serve as the starting point, no system allows the person to be introduced. If the health professional receives information (age, diagnosis, etc.) that a patient is to be admitted to a hospital ward, such information contains only objective information.

The patient's personal story can then bridge the gap between the objective perspective and a unique individual. Hence, people and lifeworld become attainable and can help understand and explain their particular illness. The story can sometimes be without words, but by meeting and communicating through the body (smiles, glances, actions), caregivers and patients can connect and build a more authentic relationship.

Similarly, a person can initiate a relationship with a beggar who is always found sitting outside the local shop and you start to worry about this person who, without verbalising it, asks for your help. The beggar's apparent suffering affects your vulnerability, causing concern and frustration. The verbal story may diminish or help comprehend these feelings because the narrative enhances understanding.

2.3 Ethics as the Basis for Health Care

A person-centred perspective on health and care uses ethics as a springboard. Such an ethical view can briefly be formulated by citing the French philosopher Paul Ricoeur: "To aim for the good life, with and for others in just institutions" [7]. This ethics guide people who often face moral dilemmas that must be resolved about ethics in every care situation or for every person seeking help. When writing about the good life, Ricoeur refers to what is good for us, namely a 'flourishing' life characterised by meaning and harmony.

Human capabilities, including those of patients, can be noted or neglected and strengthened or diminished by fellow human beings, particularly evident in situations characterised by asymmetric relationships, such as those that often occur in health care. In such cases care staff need to be aware of the importance of the relationship and how it is expressed in different situations. When the starting point is ethics and each person is understood as a unique individual, care actions will never be the same for each patient, although diagnosis and treatment are included as determining aspects.

By recognising the patients and understanding their needs and capabilities, care and treatment can be tailored to different patients and their unique needs [3, 5]. Suffering is sometimes reduced to physical or mental pain. In contrast, the most challenging suffering is a lack of recognition of a person's human capabilities, i.e. individuals are reduced to only a fraction of their potential. A capable person is vulnerable in the sense that vulnerability is not a defect in need of elimination but a constituent part of human beings that unlocks their sensitivity to the suffering of others.

2.4   Understanding the Patient's World

The jumping-off point for understanding the patient's world is that professional caregivers share the same world with the patient but that the health professional has different ways to approach and understand it. The lifeworld consists of all the immediate experiences and contacts that influence our world. Thus, all people develop their ways of understanding themselves and others, with the ultimate goal of achieving harmony and living in coexistence with others. In addition, our way of understanding things is in rapid and constant change. Accordingly, the lifeworld is a realm of both unique and shared experiences. By asking for the patients' understanding of the condition and treatment relative to their lives in general, professionals can understand what health, illness, treatment and care convey to patients and their relatives.

Our pre-understanding is based on our lifetime experiences and can be a valuable asset when we navigate the world in the presence of others. Even so, such a pre-conception can hinder further understanding as it creates certain expectations. If something is different from what we expect if something is not usual or common understanding can be greatly obscured. Everyday life would not work well and much would become complicated without the daily conversation lubricant as 'the natural setting' implies. Nevertheless, it is important to be self-critical and not take our understanding for granted, i.e. critical reflection is necessary, something that thoughtful people (such as health care practitioners) incorporate into their life. Subtle shifts in the tone of a person's voice may be observed that make us wonder if this person likes the situation as he or she claims.

Such attention that leads to reflection must always be included in health care situations. In particular, health care professionals have to listen to the premise that they do not know what the patient knows, needs, desires or is willing to disclose. Listening this way is time-consuming and a complicated process but it is essential when seeking to gain a fuller understanding of the patient's narrative. In the quest to understand the patient professionals need to be responsive, subtle and flexible.

To penetrate what strengthens a patients' vitality and help them find ways to recover an 'insider' perspective is needed [9]. This insider concept refers to gaining insight into the patient's life, where staff use their humanity to understand the patient's humanity. It is not enough to use cognitive skills. As professionals, we need emotional and bodily competence to grasp patients' deeper existential layers and their current situation. Incorporating such strategies in today's health care environment can be challenging but necessary to attain a more humanised care.

2.5   The Personal Health Plan

Establishing a health plan for care or rehabilitation requires health workers to include medical data (e.g., signs and symptoms) related to the particular condition or diagnosis. Still, the patient narratives are important in formulating such a plan but have not been given adequate space or significance in today's health care environment [10]. Patient narratives are not just one long story told by a patient on one occasion, rather, they are often a series of conversations between a patient and a health professional (perhaps with other professional or family members present) [11]. These conversations and narratives intend to clarify the context and the patient's life situation. Patients, close relatives and professionals can agree on what is relevant and what should be emphasised in the health plan. However, underscoring the value of stories is not just about verbal narratives in that patients cannot always express their story in words.

For various reasons, people may have problems expressing themselves verbally (e.g., a person with a stroke or similar disease, small children who have not yet developed their language or those who have a different first language from the caregiver). In such situations, where linguistic capacity is inadequate or non-existent, alternative strategies are needed.

As Gadamer notes, the tough thing is not finding the 'good' answers but asking the right questions [12]. The right questions are marked by the fact that we do not know the answers. We may have encountered patients with a disease that we know about as professional caregivers. However, we still lack an understanding of how the specific patient experiences the disorder, cope with daily life activities, wishes and needs. Transparency and questioning are limited by context (e.g., hospital or municipality care, emergency illness or health promotion). Consequently, there needs to be insight into how, for example, the patient's context affects both the patient and the professional's understanding of a current care situation.

2.6   An Example of a Person-Centred Care Intervention

The person-centred intervention, consisting of a combined digital platform and structured telephone support system, was provided for 6 months in addition to usual care.
The structured telephone support programme included an optional number of phone
calls with a health plan co-created and followed up by patients and health care
professionals consistent with person-centred principles. The digital platform was
built to support communication between phone calls and provide access to shared
documentation (health plans and self-ratings) and reliable information sources.

In the first telephone conversations the health care professionals encouraged the
patients to talk about their beliefs, thoughts and feelings.

The professionals established a partnership using communication skills such as
listening to participants' narratives about daily life events and the effects of their
condition. The next step entailed co-creating a health plan based on patient narratives,
including patient goals, resources and needs.

The health plan typically contained information about what they had talked about,
how the participants felt, what goals they had and what they hoped to achieve.
Participants' capacities and resources to help them achieve their respective health
goals were also included in the health plan.

Health care professionals and patients collaborated to schedule follow-up meetings. The health plan was then uploaded to the digital platform with help from health
care professionals if participants chose to write the plan themselves. The health plan
served as a leading source for impending talks and communication via the platform.
All participants and health care professionals had access to the platform during the
6-month study period. The health plan was revised during each follow-up phone call
and when needed (e.g., if the participants spontaneously contacted the health care
professionals).

The platform contained:

(1)   functionalities for two-way communication through private messages or calls,
(2)   the possibility to rate daily symptoms to be visualised as trend graphs and
(3)   an archive of the health plan.

The participants could invite and give customised access to the platform to any person
they wanted, such as informal carers, family or friends. They also could access links
to relevant websites containing information and services about their diseases. This
information was provided by patient organisations and the Swedish national support
guide (1177.se) as to an online peer-to-peer support group. A detailed description of
the intervention has been published elsewhere [13].

2.7   Evaluation of Person-Centred Care Interventions

From a person-centred perspective, health care professionals recognise patients as partners in planning and performing the care process. Moreover, person-centred care comprises shared responsibility, co-ordinated care and treatment [3, 5, 11, 14]. Early research has shown that an intervention based on person-centred principles after surgery has successfully improved daily living activities, improved care satisfaction and reduced hospital admissions [15].

Based on these findings, Ekman et al. illustrated how the ethics of person-centredness could be operationalised in practice through person-centred care, where the theoretical framework encompasses the philosophy of personhood manifested through the patient narrative, partnership and coherent documentation. One of the first controlled studies based on this framework showed reduced hospital stay for patients with CHF without worsening functional performance or increasing readmission risk [16].

Previous evaluations have reported how health professionals translate person-centred care into clinical practice and how well participants understand the established partnership and co-operation created when using this model [4, 17]. In these studies, health care professionals had to interpret how to apply person-centred care in their setting and that some aspects of the partnership created through person-centred care are not directly linked to the content of the health plan. Because a person-centred care intervention contains several interacting components, it is a complex and challenging objective [18, 19].

For example, the intervention's elements should be tailored to each participant and different clinical contexts given that the potential outcomes can be multiple and dispersed rather than linear. The design and evaluation of complex interventions need to be handled according to the complexity involved, including understanding how interventions are produced and affect participants and the settings in which they are tested and later implemented.

2.8   Effects of Person-Centred Care Measured by Controlled Studies

In a study evaluating PCC in an older patient group with CHF the length of hospital stay was reduced by 30%, activities in daily living were better preserved, uncertainty about the disease and treatment was reduced and the discharge process was more effective and less costly [16, 20].

In a randomised controlled trial, with follow-up in outpatient and primary care, PCC implementation after hospitalisation for acute coronary syndrome was evaluated. Results showed a significant, three-fold higher chance of improved self-efficacy and a return to work (or previous activity level). Moreover, the execution of PCC has proven to be efficient and cost-effective [21–23].

Another study found that using an eHealth tool combined with a person-centred approach resulted in a significant four-fold higher prospect of improved self-efficacy [24]. With regard to the ethical basis in PCC, an important finding was that patients with lower than a university education significantly improved their self-efficacy compared to those with academic degrees.

This finding confirms that person-centeredness supports equal access to care and actively reduces social disparities in health care [21]. A randomised controlled study evaluating PCC in patients with severe CHF in palliative care at home showed significant differences in reducing symptoms, increasing quality of life and decreasing rehospitalisation rates [25].

In a recent randomised controlled study person-centred care in people with chronic heart failure and/or chronic obstructive pulmonary disease was evaluated using person-centred telephone-contacts.

Results showed that only three calls were made during 6 months, on average, and self-efficacy significantly worsened in the control group and showed no change in the intervention group [26]. The significance of these studies, based on person-centred ethics and consecutive clinical trials to study practical application, is so profound that their results have been embraced by the Swedish health and social care sector and embedded in the strategic focus of the Swedish Association of Local Authorities and Regions.

The relevance and impact of the performed and supported studies have proven to be high. In addition, two EU-funded projects on PCC has been conducted.

First, within the 7th frame programme, the WE CARE project, a road map for future health in Europe, was developed by key players representing different countries and disciplines [1].

Second, a COST initiative that included 28 countries was established with three test-beds for testing and researching PCC on cost containment and quality of care in different systems [27].

Recently, a European standard has been approved for minimal patient involvement in person-centred care [28].

2.9   Conclusion

The principle of person-centred care is presented as the antithesis of reductionism. The doctrine maintains that patients are persons and should not be reduced to their disease alone. Instead, the subjectivity and integration of patients within a given environment should also be considered, including their strengths, plans and patient rights, a subset of human rights. Person-centred care implies a shift away from a model in which the patient is the passive target of medical intervention to a more contractual arrangement involving patients having an active part in their care and the decision-making process. Person-centred care includes active collaboration with the patient as a person based on the patient's narratives. Moreover, a readily accessible, tailored health plan is formulated and documented together with each patient. The main difference between person-centred care and diagnostic medicine is that the patient is accepted as a subject in person-centred care. In contrast, in conventional medicine the patient is a biological object identified by a series of diagnostic measurements. Controlled studies have found several benefits from implementing person-centred practices, including improved quality of life, self-esteem maintenance and reduced health costs.

Acknowledgements This publication is based upon work from COST Action "**European Network for cost containment and improved quality of health care-CostCares**" (CA15222), supported by COST (European Cooperation in Science and Technology)

COST (European Cooperation in Science and Technology) is a funding agency for research and innovation networks. Our Actions help connect research initiatives across Europe and enable scientists to grow their ideas by sharing them with their peers. This boosts their research, career and innovation.

https://www.cost.eu

EUROPEAN COOPERATION
IN SCIENCE & TECHNOLOGY

Funded by the Horizon 2020 Framework Programme
of the European Union

References

1. Ekman, I., Busse, R., Van Ginneken, E., Van Hoof, C., Van Ittersum, L., Klink, A., Kremer, J.A., Miraldo, M., Olauson, A., De Raedt, W., Rosen-Zvi, M., Strammiello, V., Tornell, J., Swedberg, K.: Health-care improvements in a financially constrained environment. Lancet **387**, 646–647 (2016)
2. Krause, C.: the case for quality improvement. Healthc Q **20**, 25–27 (2017)
3. Ekman, I., Hedman, H., Swedberg, K., Wallengren, C.: Commentary: Swedish initiative on person centred care. BMJ **350**, h160 (2015)
4. Britten, N., Ekman, I., Naldemirci, O., Javinger, M., Hedman, H., Wolf, A.: Learning from Gothenburg model of person centred healthcare. BMJ **370**, m2738 (2020)
5. Ekman, I., Swedberg, K., Taft, C., Lindseth, A., Norberg, A., Brink, E., Carlsson, J., Dahlin-Ivanoff, S., Johansson, I.L., Kjellgren, K., Liden, E., Ohlen, J., Olsson, L.E., Rosen, H., Rydmark, M., Sunnerhagen, K.S.: Person-centered care–ready for prime time. Eur. J. Cardiovasc. Nurs. **10**, 248–251 (2011)
6. Nussbaum, M.: Creating capabilities: the human development approach. The Belknap Press of Harvard University Press, Cambridge, Massachusetts and London (2011)
7. Ricoeur, P.: Oneself as another. The University of Chicago Press, Chicago and London (1992)
8. Sen, A.: Capability and well-being. In: Nussbaum, M., Sen, A. (eds.) The Quality of Life. Oxford: Clarendon Press (1993)
9. Todres, L., Galvin, K.T., Dahlberg, K.: "Caring for insiderness": phenomenologically informed insights that can guide practice. Int. J. Qual. Stud. Health Well-Being **9**, 21421 (2014)
10. Wallstrom, S., Ekman, I.: Person-centred care in clinical assessment. Eur. J. Cardiovasc. Nurs. 1474515118758139 (2018)
11. Coulter, A., Entwistle, V. A., Eccles, A., Ryan, S., Shepperd, S., Perera, R.: Personalised care planning for adults with chronic or long-term health conditions. Cochrane Database Sys. Rev. **3**, CD010523 (2015)
12. Gadamer, H.-G.: Truth and method, 2nd edn. The Continuum Publishing Company, New York (1995)

13. Ali, L., Wallström, S., Barenfeld, E., Fors, A., Fredholm, E., Gyllensten, H., Swedberg, K., Ekman, I.: Protocol: person-centred care by a combined digital platform and structured telephone support for people with chronic obstructive pulmonary disease and/or chronic heart failure: study protocol for the PROTECT randomised controlled trial. BMJ Open **10** (2020)

14. Hakansson Eklund, J., Holmstrom, I.K., Kumlin, T., Kaminsky, E., Skoglund, K., Hoglander, J., Sundler, A.J., Conden, E., Summer Meranius, M.: "Same same or different?" A review of reviews of person-centered and patient-centered care. Patient Educ. Couns. **102**, 3–11 (2019)

15. Olsson, L.E., Karlsson, J., Ekman, I.: The integrated care pathway reduced the number of hospital days by half: a prospective comparative study of patients with acute hip fracture. J. Orthop. Surg. Res. **1**, 3 (2006)

16. Ekman, I., Wolf, A., Olsson, L.E., Taft, C., Dudas, K., Schaufelberger, M., Swedberg, K.: Effects of person-centred care in patients with chronic heart failure: the PCC-HF study. Eur. Heart J. **33**, 1112–1119 (2012)

17. Wolf, A., Moore, L., Lydahl, D., Naldemirci, O., Elam, M., Britten, N.: The realities of partnership in person-centred care: a qualitative interview study with patients and professionals. BMJ Open **7**, e016491 (2017)

18. Craig, P.D.P., Macintyre, M., Michie, S., Nazareth, I., Petticrew, M.: Developing and Evaluating Complex Interventions: New Guidance. London: Medical Research Council (2006)

19. Craig, P., Dieppe, P., Macintyre, S., Michie, S., Nazareth, I., Petticrew, M.: Medical research council, G. Developing and evaluating complex interventions: the new Medical Research Council guidance. BMJ **337**, a1655 (2008)

20. Hansson, E., Ekman, I., Swedberg, K., Wolf, A., Dudas, K., Ehlers, L., Olsson, L.E.: Person-centred care for patients with chronic heart failure—a cost-utility analysis. Eur. J. Cardiovasc. Nurs. **15**, 276–284 (2016)

21. Fors, A., Gyllensten, H., Swedberg, K., Ekman, I.: Effectiveness of person-centred care after acute coronary syndrome in relation to educational level: subgroup analysis of a two-armed randomised controlled trial. Int. J. Cardiol. **221**, 957–962 (2016)

22. Pirhonen, L., Olofsson, E.H., Fors, A., Ekman, I., Bolin, K.: Effects of person-centred care on health outcomes—a randomized controlled trial in patients with acute coronary syndrome. Health Policy **121**, 169–179 (2017)

23. Pirhonen, L., Bolin, K., Olofsson, E.H., Fors, A., Ekman, I., Swedberg, K., Gyllensten, H.: Person-centred care in patients with acute coronary syndrome: cost-effectiveness analysis alongside a randomised controlled trial. Pharmacoecon Open **3**, 495–504 (2019)

24. Wolf, A., Fors, A., Ulin, K., Thorn, J., Swedberg, K., Ekman, I.: An eHealth diary and symptom-tracking tool combined with person-centered care for improving self-efficacy after a diagnosis of acute coronary syndrome: a substudy of a randomized controlled trial. J. Med. Internet Res. **18**, e40 (2016)

25. Brannstrom, M., Boman, K.: Effects of person-centred and integrated chronic heart failure and palliative home care. PREFER: a randomized controlled study. Eur. J. Heart Fail (2014)

26. Fors, A., Blanck, E., Ali, L., Ekberg-Jansson, A., Fu, M., Lindstrom Kjellberg, I., Makitalo, A., Swedberg, K., Taft, C., Ekman, I.: Effects of a person-centred telephone-support in patients with chronic obstructive pulmonary disease and/or chronic heart failure—a randomized controlled trial. PLoS One **13**, e0203031 (2018)

27. Costcares: Person centred care 4 sustainable health systems [Online]. Brussells: European Union (2021). https://costcares.eu/

28. Cen/Tc450: Patient involvement in person-centred care [Online]. Brussells (2020). https://standards.cen.eu/dyn/www/f?p=204:110:0::::FSP_PROJECT,FSP_LANG_ID:65031,25&cs=170DF874D1EFAE86E8A8205451C28A49D

Chapter 3
Person-Centred Care Implementation: Design and Evaluation Considerations

Heather L. Rogers, Vítor Raposo, Maja Vajagic, and Bojana Knezevic

Abstract The Gothenburg model of Person-Centred Care (PCC) is an evidence-based intervention shown to improve care and health outcomes while maintaining cost. Other health systems could benefit from its sustainable implementation. The WE-CARE implementation framework, adapted by COSTCares, provides a base set of enablers and outcomes recommended for the design and evaluation of PCC. The methodology is extended using implementation science to systematically address contextual factors at different levels. Evidence-based frameworks, such as the Consolidated Framework for Implementation Research (CFIR), for example, and hybrid effectiveness-implementation study designs can be used. Additional enablers to consider when designing and evaluating PCC implementation strategies are discussed. The outcomes of quality of care and cost can be addressed using a Value for Money (VfM) framework. Various VfM methods and analysis models can be incorporated into PCC implementation research design in order to influence policy makers and health system decision makers towards the sustainable uptake of PCC.

Keywords Person-centred care · Implementation science · Value for Money frameworks

---

H. L. Rogers (✉)
Biocruces Bizkaia Health Research Institute, Barakaldo, Spain

IKERBASQUE Basque Foundation for Science, Bilbao, Spain

V. Raposo
Faculty of Economics, Centre for Business and Economics Research (CeBER), Centre of Health Studies and Research of the University of Coimbra, University of Coimbra, Av. Dr. Dias da Silva 165, 3004-512 Coimbra, Portugal
e-mail: vraposo@fe.uc.pt

M. Vajagic
Croatian Health Insurance Fund, Zagreb, Croatia
e-mail: maja.vajagic@hzzo.hr

B. Knezevic
Faculty of Kinesiology, University Hospital Centre Zagreb, University of Zagreb, Zagreb, Croatia
e-mail: bojana.knezevic@kbc-zagreb.hr

© The Author(s) 2022
D. Kriksciuniene and V. Sakalauskas (eds.), *Intelligent Systems for Sustainable Person-Centered Healthcare*, Intelligent Systems Reference Library 205,
https://doi.org/10.1007/978-3-030-79353-1_3

3.1   Introduction

Demographic changes and the rise of chronic diseases have led to increased demand for healthcare and are some of the many challenges currently facing health and social care systems. Viable solutions that maintain or decrease cost while improving quality of care are required. Person-centred care (PCC) has been identified as a possible intervention to improve care and maintain or reduce healthcare costs, especially for patients with chronic and long-term conditions [15]. This chapter addresses evaluation and design considerations for the sustainable implementation of PCC. First, the intervention and an implementation framework of enablers and outcomes will be presented. Then additional enablers will be detailed, along with evidence-based considerations from implementation science regarding assessment of contextual factors and study designs for PCC implementation. Finally, tools and methodologies to assess the cost-effectiveness aspects of the outcomes using a Value for Money framework will be described.

3.2   The PCC Intervention

The Gothenburg model of PCC incorporates an ethos that recognizes the person behind the patient presenting with a disease [24]. It has a basis in the ethical perspective of care provider as healer who assists identification of the patient's strengths and resources to facilitate healing when facing a diagnosis that threatens the self [28]. PCC uses the illness narrative as the foundation for collaborative, equalitarian relationships between care providers and the patient/person expert [5]. Furthermore, PCC endorses and promotes egalitarian principles [34].

In practice, the Gothenburg model of PCC involves three "routines" to help care providers systematically and consistently implement PCC. These routines, or pillars, help providers put the person before the disease in their interactions with patients in the context of their hectic, day-to-day schedules. According to Ekman and colleagues [11], the three routines are as follows:

1. **Initiate the partnership:** The providers invite the patient to relate a narrative account of his/her experience with the disease or condition. The patient provides a personal account of his/her illness, symptoms, and impact on everyday life. The person's beliefs, feelings, and preferences are expressed and heard by the providers via the narrative. The person's resources and strengths are identified and assessed. These facets are then leveraged to enhance self-management.

2. **Working the partnership**: The provider-person interactions and patient narrative are used as a basis for care planning. The patient joins the care team as an expert in his/her own life. Through discussion, all the providers and the person—and caregivers, as appropriate—engage in deliberative shared-decision making (see details in Chap. 5). Consensus is reached on care goals and all possible

options are assessed in this care team to ensure selection according to patient values, beliefs, preferences, values, lifestyle, and health issues.

3. **Safeguarding the partnership:** Through documentation, the providers validate the patient's preferences, values, and beliefs and involvement as an expert member of the care team. The PCC process, including the patient narrative and care plan, is documented for the care team. The care team meets regularly to review progress and required adaptations. In this way, continuity of PCC is ensured.

3.3 COSTCares PCC Implementation Framework

Although PCC is implemented at the micro-level by care providers, PCC is also implemented in the meso- (e.g., organizational) and macro- (e.g., policy and financing) levels of service delivery within a health system. PCC implementation involves education of health professionals, as well as constant communication and collaboration among health workers and patients [15]. Because PCC often requires a change in current processes of patient care and involves teams of professionals, PCC may affect many aspects of care—functional (e.g., support functions like financial management), organizational (e.g., networks), professional (e.g., alliances of professionals) and clinical (e.g., processes). At the system level, horizontal (across the same level of care provision) and vertical (primary-secondary-tertiary care) integration may be needed. The degree of PCC integration may also differ depending on context, ranging from separate linked structures, to coordination of care, to full integration of PCC. Clear goal definition and examination of implementation context is important before determining how PCC might be sustainably implemented and assessed [3].

The European Commission (EC) funded WE-CARE consortium developed a model positing five critical macro-level enablers for sustainable PCC leading to quality of care and cost containment as outcomes [12]. The enablers include information technology, quality measures, infrastructure, incentive systems and contracting strategies. COSTCares, COST Action 15,222, was a follow-on EC project, which extended this framework by adding an additional enabler: cultural change [21]. These enablers overlap and interact. They are hypothesized to help to drive the uptake, adoption, and maintenance of PCC and therefore influence the two outcomes of quality of care and cost containment. Each of these enablers and outcomes, at minimum, should be considered when designing and evaluating implementation strategies.

Examination of additional enablers of PCC implementation may be warranted. For instance, a review of integrated care experiences in Europe by a European Commission Expert Group on Health Systems Performance [3] identified macro- and meso-level factors influencing successful implementation that likely apply to PCC implementation. The WE-CARE enablers were included (e.g., organizational change, financing and incentives, information communication technology infrastructure and solutions, and monitoring/evaluation system), and the following other factors were important: Political support and commitment, Governance, Stakeholder engagement, Leadership, Collaboration and trust, and Workforce education and training.

Santana and colleagues [29] propose valuable a general conceptual framework or roadmap to guide systems and organizations in PCC provision and evaluation and quality of care improvement in general. The roadmap is based on the Donabedian model domains of structure, process and outcome. The emphasis is on the structure, or health system domain as it provides the context of care delivery. The most important constructs in this domain are the PCC culture in the continuum of care, the educational programs of health workers, a supportive environment, the development of supportive health technologies, the monitoring and measurement of PCC performance, and feedback from patients. Process constructs in the model include communication and interaction between the patients and providers of care, respectful and compassionate care, and including patients as partners in care. Outcome constructs in the model relate to the results from the integration of PCC care across the health and social care system, impact on health and social care professionals and patients, and the value of PCC implementation [29].

3.4 Implementation Science Frameworks to Examine Contextual Factors

The field of implementation science was developed to address the research-to-practice gap. Implementation science is defined as "The scientific study of methods to promote the systematic uptake of research findings and other evidence-based practices into routine practice, and, hence, to improve the quality and effectiveness of health services and care" [10]. While clinical research addresses the "what", implementation research addresses the "how". Nilsen [25] offers a schematic to organize the theoretical approaches used by implementation science. Each approach is categorized based on function and include:

1. to describe or guide the process of translating research into practice (e.g., via process models);
2. to understand or explain what influences implementation outcomes, for example via determinant frameworks, classic theories, or implementation theories; or
3. to evaluate implementation, using evaluation frameworks.

One of many comprehensive models that can be used for all three functions is the Consolidated Framework for Implementation Research (CFIR; [8]). It consists of five domains, including the intervention itself with adaptations, outer setting, inner setting, characteristics of the individuals involved, and the implementation process. Each domain consists of various constructs. There is a website (https://cfirguide. org/) with supporting materials, including a detailed description of each construct, a qualitative codebook with operational definitions and inclusion and exclusion criteria for each mutually exclusive construct, and quantitative measures as they become available.

Another practice resource for implementation, which cites the CFIR and other valuable frameworks for design and evaluation is ImpRes—Implementation Science Research Development Tool which can be found at: https://impsci.tracs.unc.edu/wp-content/uploads/ImpRes-Guide.pdf [16].

In the field of implementation science, the intervention is studied separately from the implementation strategies used to facilitate its sustainable uptake. Implementation study designs examine intervention-implementation effectiveness in real-life conditions. This is in contrast to pharmacological research, in which the gold standard to determine drug effectiveness is the randomized controlled trial. In implementation research, hybrid effectiveness-implementation designs assess both the clinical effectiveness of an intervention and its implementation. Context of intervention implementation (e.g., enablers) are considered a priori. Curran and colleagues [7] proposed three general groups of hybrid effectiveness-implementation designs, which differ depending on whether the intervention or the implementation is the primary focus:

1. Hybrid Effectiveness-Implementation Type I—Intervention primary: Examines the effects of an intervention on relevant outcomes while observing and gathering information on implementation;
2. Hybrid Effectiveness-Implementation Type II—Both the intervention and its implementation are primary: Simultaneously examines the intervention and implementation strategies; and
3. Hybrid Effectiveness-Implementation Type III—Implementation primary: Examines the effects of an implementation strategy on an intervention while observing and gathering information on the intervention's impact on relevant outcomes.

Lane-Fall and colleagues [19] offer a visual representation of intervention and hybrid effectiveness-implementation designs in the form of a decision tree. They call it the "subway line of translational research". The first question to determine which line one should take is if the intervention has shown efficacy. If not, efficacy research into the intervention, such as laboratory studies, are required. If the answer is yes, the next question is if the intervention has shown effectiveness in different contexts. If not, effectiveness research, such as randomized controlled trials, is warranted. If effectiveness is not fully demonstrated, hybrid effectiveness-implementation trials are considered valuable. If an intervention has shown effectiveness, hybrid effectiveness-implementation trials are also used, with a focus on implementation. Three study designs are proposed in order of advancing complexity: quantitative–qualitative (mixed) methods studies to better understand the role of contextual factors on intervention outcomes, designing implementation strategies, and finally testing and evaluating the effectiveness of different implementation strategies. A final stop that was not addressed by Lane-Fall et al. [19] would be the optimization of successful implementation strategies. This stepwise process and use of hybrid effectiveness-implementation research designs helps to understand the role of context in intervention outcomes. This knowledge is not only useful to ensure the best possible outcomes in a specific setting, but it also facilitates the adaptation and transfer of interventions to other settings and/or the scaling-up of interventions to new, larger settings.

3.5 Examination and Evaluation of PCC in Context: Additional Considerations

Many conceptual frameworks for implementation evaluation are available, but it is important to develop practical guidelines for implementation. In this respect, logic models are an effective tool for intervention design, implementation, and evaluation.

A logic model is a visual representation of a theory of action or program logic, or program theory. It provides a simplified picture of the relationships between the intervention inputs (resources, strategies, activities) and the desired outcomes of the program. As part of COSTCares Working Group 3, Lloyd and colleagues [21] used program theory to create if–then statements to hypothesize how the different enablers might interact to facilitate PCC intervention adoption in order to generate the goals of cost containment and better quality of care. Each enabler was considered on micro, meso and/or macro levels. In brief, conclusions are offered regarding recommendations for PCC evaluation in practice [21]. For instance:

1. Health outcomes measures should be relevant to patients and their families, as well as health care workers and decision makers. Health outcomes measures should include the patient experience and the markers of quality.
2. The main areas of implementation evaluation of PCC from patient point of view are functional ability, care experience, self-efficacy and cost of care.
3. Pre- and post-intervention/implementation data collection points are important.
4. Continuous monitoring, with feedback to stakeholders involved, and over a long follow-up period is key.
5. It is crucial that the evaluation measures reflect PCC goals and results regarding goal achievement. These measures must be accurate, objective and verifiable, and indicative of real performance.
6. A minimum data set of PCC-related indicators should be compounded. These may be composed of routinely collected data, questionnaire data and qualitative data, as well as results of outcomes concerning health, quality and cost.

Santana and colleagues [30] concur that the definition of PCC quality indicators is not clear and there is a lack of current indicators available to assess implementation in the care setting.

The selection of appropriate indicators to measure enablers, the PCC intervention, and outcomes requires careful thought. There are several criteria that can be used to determine which indicators might be most appropriate. The European Commission Expert Group on Health System Performance recommend that the most feasible indicators be chosen, for instance those that are already in use with logistics in place (such as existing datasets) for data collection [3]. The Organisation for Economic Co-operation and Development (OECD) Health Care Quality Indicators Project suggest the following criteria for the selection of indicators: validity, reliability, relevance, action ability, feasibility, and comparability [4]. Raleigh and colleagues [27] add to this list: accuracy, meaningfulness, and avoidance of perverse incentives. These last criteria are particularly important in the evaluation of PCC.

The following aspects must be operationalized when undertaking the implementation of PCC in a health system. What are the main aims of PCC? What are the desired outcomes of PCC? What are the time frames over which the outcomes could be achieved? What are the best possible outcomes? What might be some unintended effects? What is the scope of implementation? What setting or settings will be addressed? Will there be inter-sectorial collaboration (e.g., between the health sector, social/community care sector, education sector)? How can the impact be measured? To what extent can a given measure meet the indicator criteria defined above? What data are already available that might be relevant? What are the pros and cons of those readily available indicators? What are the options for new and innovative ways to collect indicator data?

New indicators of PCC may need to be developed that evaluate structure (e.g., assessment of the basic conditions and system levers needed for transformation), process (e.g., focusing on the areas where there are more barriers), and outcomes (especially concerning the patient experience). All indicators must support evidence-based investment and the impact of every change must be monitored and evaluated comprehensively from different perspectives—for instance by patients, family/caregivers, health and care workers, administrators [3]. Since the impact of PCC on outcomes may only become apparent in the longer-term, an emphasis on shorter-term intermediate outcomes and process indicators is warranted. However, measurement of longer-term outcomes (especially health outcomes) is needed in order to collect potentially convincing evidence as to cost-effectiveness.

Gyllensten and colleagues [15] support the development of a core outcome set for PCC evaluation. The set of outcomes should include economic, clinical, humanistic, and unintended outcomes, as well as measures of patient experiences with healthcare services. Given the importance of the cost containment and quality of care outcomes to the PCC Implementation Framework, the rest of this chapter is dedicated to these considerations.

3.6 Value for Money and Economic Evaluation Tools

Value for money (VfM) and economic evaluation tools have been central to health policy decisions, accountability, healthcare delivery and healthcare systems [22, 26, 32]. Economic evaluation helps identify the more relevant alternatives, allows different analysis viewpoints, raises quantification over the informal assessment, and increases explicitness and accountability in decision-making [9, 14].

Governments are increasingly required to strategically manage scarce resources by investing in services that provide the best health outcomes [33]. The rapid diffusion of health technologies brings increased challenges to provide high quality and innovative care to meet population health needs most effectively while managing constrained healthcare budgets and safeguarding equity, access and choice [2, 22, 26, 33]. As argued by Drummond et al. [9], whatever the context or specific decision, a common question is posed: Are we satisfied that the additional health care resources (required to make the procedure, service, or programme available to those who could benefit from it) should be spent in this way rather than some other ways?

Another important reason for the importance of VfM and economic evaluation relates to accountability assuring that taxpayers and founders money is being spent wisely, and reassuring healthcare users and other stakeholders that their claims and interests on the health system are being treated fairly and consistently [13, 18, 32, 33].

VfM includes the three E's in its assessments [13]:

1. Economy (minimizing the cost of inputs, while bearing in mind quality),
2. Efficiency (achieving the best rate of conversion of inputs into outputs, while taking in mind quality), and
3. Effectiveness (achieving the best possible result for the level of investment, while maintaining in mind equity).
4. A fourth E was added to considerer Equity, ensuring that benefits are distributed fairly [17].

Figure 3.1 presents VfM main framework, their components and the relations between the four E's.

According to Smith [32], several components of VfM need to be considered when developing any VfM measure: eventual outcomes of interest, intermediate outputs and activities, inputs, possible external constraints on achieving VfM, and whether a long or short time horizon is being adopted. Outcomes are the valued outputs that usually are grouped on four broad categories: health gains, the patient experience, inequalities, and the broader social and economic benefits of health services.

Fleming [13] identifies six main methods that can be used to assess VfM: Cost-Effectiveness Analysis (CE analysis), Cost-Utility Analysis (CU analysis), Cost-Benefit Analysis, Social Return on Investment (SROI), Rank correlation of cost versus impact, and Basic Efficiency Resource Analysis (BER analysis). Table 3.1 presents a brief description of each method associated with VfM and when each should be applied.

The purpose of economic evaluation is to inform decisions. It deals with both inputs and outputs (costs and consequences) of alternative courses of action and is concerned with choices. Decision-makers face the problem of scarce resources (people, time, facilities, equipment and knowledge) and since the effects of choosing one course of action over another will not only have effects on health but also on health care resources as well as other effects outside health care, informing health

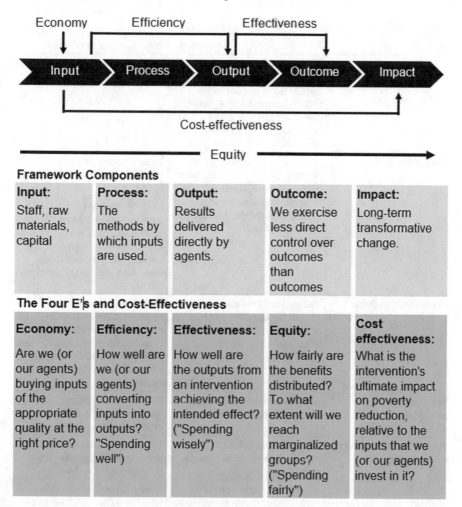

Fig. 3.1 Value for Money framework (Adapted from [18])

care decisions requires consideration of costs and benefits. Table 3.2 synthesizes additional information about five types of studies analyzing costs.

The costs involved in PCC implementation must be calculated, and all of the previous methods can be used to determine the return of investment or the value for money for private or public financing of PCC.

Table 3.1 Methods for evaluating VfM (Adapted from [13])

Method	Brief description of the method	When to apply the method
Cost-effectiveness analysis	The evaluation of two or more alternatives, based on the relative costs and outcomes (effects), in reaching a particular goal. This method can be used when comparing programs that aim to achieve the same goal	Comparing programmes that aim to achieve the same goal
Cost-utility analysis	The evaluation of two or more alternatives by comparing their costs to their utility or value (a measure of effectiveness developed from the preferences of individuals)	Used where monetizing outcomes is not possible or appropriate
		Most commonly used in health through quality adjusted life years (QALY). The QALY allows the comparison of medical interventions by the number of years that they extend life
Cost-benefit analysis	The evaluation of alternatives by identifying the costs and benefits of each alternative in money terms and adjusting for time	Used to identify if a course of action is worthwhile in an absolute sense—whether the costs outweigh the benefits—and allows for comparison among alternatives that do not share the same objective or the same sector
Social return on investment	Measures social, environmental and economic costs and benefits	Used when comparing programmes with different goals or in different sectors
Rank correlation of cost versus impact	Allows for the relative measurement of VfM across a portfolio of initiatives	Used to rank and correlate costs and impact of different programmes or initiatives
Basic efficiency resource analysis	Provides a framework for evaluating complex programmes by comparing impact to resources and offering a relative perspective on performance where units analyzed are judged in comparison to other peer units	Used to examine the relative value on a four-quadrant graph based on costs and impacts

3.7 Inform Decisions and Justify the Value—Inputs and Outputs

Most of the methods presented previously are concerned with choices when comparing costs and consequences (economic, clinic and humanistic outcomes). For instance, the ECHO model [6] incorporates costs, economic outcomes, and interrelationships with the clinical and humanistic outcomes. The same arguments are used

Table 3.2 Economic evaluation using costs (Adapted from [9])

Type of study	Costs measurement	Consequences	
		Indentification	Measurement
Cost analysis	Monetary units	Not considered. The consequences are common to all considered alternatives	
Cost-effectiveness analysis	Monetary units	Single effect of interest, common to both alternatives, but achieved to different degrees	Natural units (e.g. life-years gained, disability days saved, blood pressure reduction, levels of LDL/HDL, etc.)
Cost-utility analysis	Monetary units	Single or multiple effects, not necessarily common to both alternatives	Healthy years, typically measured as QUALYs (quality-adjusted life-years)
Cost–benefit analysis	Monetary units	Single or multiple effects, not necessarily common to both alternatives	Monetary units

by others when discussing VfM in healthcare [13, 22, 32, 33] or healthcare economic evaluation [9, 14]. That is why these models are valuable to use in the evaluation of PCC implementation. Some specific considerations regarding the costs involved and the consequences of PCC implementation are described below.

3.7.1 Costs

VfM and the economic evaluation literature can give an essential contribution to identifying the main costs involved in PCC implementation. For example, Gold et al. [14], more focused on a cost-effectiveness analysis, identifies costs related with changes in the use of healthcare resources, changes in the use of non-healthcare resources, changes in the use of informal caregiver time and changes in the use of patient time (for treatment). Similarly, Drummond et al. [9], in a broader perspective of economic evaluation involving costs and different types of analysis, identifies health sector costs, other sector costs, patient/family costs, and productivity losses. Table 3.3 provides a high-level overview of these two models.

According to Gold et al. [14], direct health care costs include all types of resource use, including the consumption of professional, family, volunteer, or patient time and the costs of tests, drugs, supplies, healthcare personnel, and medical facilities. Non-direct health care costs include the additional costs related with the intervention, such as those for childcare (for a parent attending a treatment), the increase of costs required by a dietary prescription, and the costs of transportation to and from the health facilities, they also include the time family, or time volunteers spend providing

Table 3.3 Cost components

Costs [14]	Costs [9]
Changes in the use of healthcare resources	Health sector costs
Changes in use of non-healthcare resources	Other sector costs
Changes in the use of informal caregiver time	Patient/family costs
Changes in the use of patient time (for treatment)	Productivity losses

home care. Patient time costs include the time a person spends seeking care or participating in or undergoing intervention or treatment. Relevant time costs include travel and waiting time as well the time receiving treatment.

Drummond et al. [9] indicate that health sector costs can be variable (such as the time of health professionals or supplies) and fixed or overhead costs (such as light, heat, rent, or capital costs). The other sector costs refer to consumed resources from other public agencies or the voluntary sector. Person/family costs refer to any out-of-pocket expenses incurred by patients or family members as well as the value of any resources that they contribute to the treatment process. Productivity costs include (1) the costs associated with lost or impaired ability to work or to engage in leisure activities due to morbidity and (2) lost economic productivity due to death.

According to the metrics framework described by Lloyd and colleagues (2020) from the COSTCares Working Group 3, it is possible to calculate some of the leading direct healthcare costs involved with administrative data extraction for PCC monitoring and evaluation. The calculation of other costs (mainly direct non-healthcare costs, patient times costs and productivity costs) requires the use of different tools and data sources to get the accurate values (or the closest approximations possible) to be considered in the evaluation analysis [21].

3.7.2 Valuing Healthcare and Health Effects

The prime objective of healthcare is to improve health, and different categories of goals and outcomes can be identified in this respect. According to Smith [32], responsiveness to patients' needs, addressing inequalities, and broader economic objectives are the leading healthcare goals. Many treatments offer broader social and economic benefits to patients, families and society. Other authors, like Cheng et al. [6], focus on economic outcomes and their interrelationships with the clinical and humanistic outcomes.

Drummond et al. [9] sustain that the literature on economic evaluations contains studies using several types of outcome measures: clinical outcomes, quality of life measures, and generic measures of health gain like Quality-adjusted life years

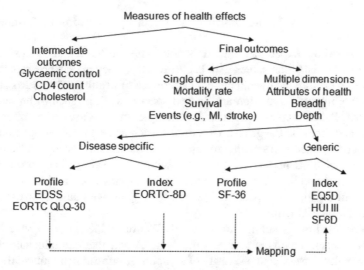

Fig. 3.2 A taxonomy of measures of health effects (Adapted from [9])

(QALYs), the Disability-Adjusted Life-Year (DALY), Short Form-36 (SF-36 Quality of Life measure), EQ5D (a family of instruments to describe and value health from EuroQol), and SF6D (Short-Form Six-Dimension, a health index designed for calculating QALYs). A taxonomy of the alternative measures of health effect is given in Fig. 3.2.

Clinical outcomes are the most common health outcome category to be considered in clinical trials and observational studies. They would be captured in hybrid implementation-effectiveness designs. Some economic evaluations use these outcomes, as reported in the relevant study of the PCC intervention, as the measure of health gain.

Humanistic outcomes are outcomes based on a patient's perspective (e.g. patient-reported scales that indicate pain level, degree of functioning). In this category, there are health-related quality of life (HRQoL) and the range of measures collectively described as patient-reported outcomes (PROs). Patient-reported outcomes (PROs), or patient-reported outcomes measures (PROMs), are information provided by the patient about their symptoms, quality of life, adherence, or overall satisfaction [23]. PROs refer to patient ratings about several outcomes, including health status, health-related quality of life, symptoms, functioning, satisfaction with care, and treatment satisfaction. The patient can also report about their health behaviors, including adherence and well-being habits.

Generic PRO questionnaires are measurement instruments designed to be used across different subgroups of individuals and contain common domains that are relevant to almost all populations. Examples of a generic PRO measure are the Sickness Impact Profile (SIP) or SF-36 that measures general health perception, pain, physical functioning, role functioning, social functioning, mental health and vitality. An instrument that assesses a more restricted set of domains is the Index of Activities of Daily Living, which measures independence in performing basic functioning.

Patient-reported experience measures (PREMs) are tools and instruments that report patient satisfaction scores with health service. They are generic tools that are often used to capture the overall patient experience of health care. PREMs are often used on the broader population and in non-specific settings such as an outpatient department. Patient experience tools, for example, may be used to monitor patient feedback and focus on the general experience, such as customer service rather than an experience related to a specific disease. These instruments or tools have revealed positive associations between patient satisfaction and safety. They are a reliable measure of how well a hospital or other health unities can provide good quality service from a patient perspective. Therefore, they are well-suited for use in PCC implementation evaluation.

Time devoted to collecting PROs and PREMs turns into invested time that can benefit the person receiving care and the organization that can allocate resources more optimally. Assessing the severity of symptoms, informing treatment decisions, tracking outcomes, prioritize patient-provider discussions, monitoring general health and well-being, and connecting providers to patient-generated health data are PROs and represent different ways of creating value [20].

The International Consortium for Health Outcomes Measurement (ICHOM) collaborates with patients and healthcare professionals to define and measure patient-reported outcomes to improve quality of care and value. The ICHOM website (https://www.ichom.org) is a valuable resource, as it is possible to find several standardized outcomes, measurement tools and time points and risk adjustment factors for a given condition that could be used in PCC implementation evaluation. The Patient-Reported Outcomes Measurement Information System (PROMIS; https://www.healthmeasures.net), the Outcome Measures in Rheumatology (OMERACT; https://omeract.org) and the Consensus-based Standards for the Selection of Health Measurement (COSMIN; https://www.cosmin.nl) are other useful references for these kinds of patient-reported instruments.

A more recent source of data is patient-reported information (PRI) proposed by Baldwin et al. [1]. According to those authors, PRIs take up the PRO tool and reinforce the patient perspective. This new perspective is related to social networking that enables patients to publish and receive communications very quickly. Many stakeholders, including patients, are using these media to find new ways to make sense of diseases, to find and discuss treatments, and to give support to patients and their caregivers.

According to Schlesinger et al. [31], PRI pinpoints the limits of traditional measurement techniques to incorporate narrative components into the evaluation and can be used to improve clinical practice. Those authors identify four forms of PRIs:

1. patient-reported outcomes measuring self-assessed physical and mental well-being,
2. surveys of patient experience with clinicians and staff,

3. narrative accounts describing encounters with clinicians in patients own words, and
4. complaints/grievances signalling patients distress when treatment or outcomes fall short of expectations.

The narrative aspects of PRIs align well with PCC, and more research is needed to uncover the value of these data sources for PCC implementation evaluations in the future.

3.8 Conclusion

In summary, the Gothenburg model of PCC is an evidence-based intervention shown to improve care and health outcomes while maintaining cost. Other health systems could benefit from its sustainable implementation. The WE-CARE implementation framework, adapted by COSTCares Working Group 3, provides a base set of enablers and outcomes recommended for the design and evaluation of PCC. These core enablers of information technology, quality measures, infrastructure, incentive systems, contracting strategies, and cultural change may need to be broadened, depending on the health system. Political support and commitment, governance, stakeholder engagement, leadership, collaboration and trust, and workforce education and training are additional enablers that warrant consideration for sustainable implementation of PCC. Contextual factors, and the interaction of enablers, must be examined. Implementation science offers evidence-based frameworks to systematically evaluate factors that will influence successful uptake of PCC. Regarding the outcomes of quality of care and costs, a Value for Money framework, along with associated cost-effectiveness methods and analysis models, are other important aspects of PCC evaluation and design. PROs, PROMs, PREMs, and especially PRIs should be considered as means to capture the perspective of the patient as a person, which is at the center of PCC. In conclusion, comprehensive assessments of PCC, enablers, and outcomes should be incorporated into PCC implementation design in order to influence policy makers and health system decision makers towards the sustainable uptake of PCC.

Acknowledgements This publication is based upon work from COST Action "**European Network for cost containment and improved quality of health care-CostCares**" (CA15222), supported by COST (European Cooperation in Science and Technology)

COST (European Cooperation in Science and Technology) is a funding agency for research and innovation networks. Our Actions help connect research initiatives across Europe and enable scientists to grow their ideas by sharing them with their peers. This boosts their research, career and innovation.

https://www.cost.eu

References

1. Baldwin, M., Spong, A., Doward, L., Gnanasakthy, A.: Patient-reported outcomes, patient-reported information. Patient: Patient-Cent. Outcomes Res. **4**(1), 11–17 (2011)
2. Banke-Thomas, A., Nieuwenhuis, S., Ologun, A., Mortimore, G., Mpakateni, M.: Embedding value-for-money in practice: a case study of a health pooled fund programme implemented in conflict-affected South Sudan. Eval. Program Plan. **77**, 101725 (2019). https://doi.org/10.1016/j.evalprogplan.2019.101725
3. Building Blocks: Tools and methodologies to assess integrated care in Europe. Report by the Expert Group on Health Systems Performance Assessment. European Union, Luxembourg (2017). Available at: https://ec.europa.eu/health/sites/health/files/systems_performance_asse ssment/docs/2017_blocks_en_0.pdf
4. Carinci, F., Van Gool, K., Mainz, J., et al.: Towards actionable international comparisons of health system performance: expert revision of the OECD framework and quality indicators. Int. J. Qual. Health Care: J. Int. Soc. Qual. Health Care/ISQua **27**, 137–146 (2015)
5. Charon R.: The patient–physician relationship. Narrative medicine: a model for empathy, reflection, profession, and trust. JAMA **286**, 1897–902 (2001)
6. Cheng, Y., Raisch, D.W., Borrego, M.E., Gupchup, G.V.: Economic, clinical, and humanistic outcomes (ECHOs) of pharmaceutical care services for minority patients: a literature review. Res. Soc. Adm. Pharm. **9**(3), 311–329 (2013)
7. Curran, G.M., Bauer, M., Mittman, B., Pyne, J.M., Stetler, C.: Effectiveness-implementation hybrid designs: combining elements of clinical effectiveness and implementation research to enhance public health impact. Med. Care **50**, 217–226 (2012)
8. Damschroder, L.J., Aron, D.C., Keith, R.E., Kirsh, S.R., Alexander, J.A., Lowery, J.C.: Fostering implementation of health services research findings into practice: a consolidated framework for advancing implementation science. Implement. Sci. **4**, 50 (2009)
9. Drummond, M.F., Sculpher, M.J., Claxton, K., Stoddart, G.L., Torrance, G.W.: Methods for the Economic Evaluation of Health Care Programmes. Oxford University Press, Oxford (2015)
10. Eccles, M.P., Mittman, B.S.: Welcome to implementation science. Implement. Sci. **1**, 1 (2006)
11. Ekman, I., Swedberg, K., Taft, C., Lindseth, A., Norberg, A., Brink, E., Carlsson, J., Dahlin-Ivanoff, S., Johansson, I.L., Kjellgren, K., Lidén, E., Öhlén, J., Olsson, L.E., Rosén, H., Rydmark, M., Stibrant Sunnerhagen, K.: Person-centered care — Ready for prime time. Eur. J. Cardiovasc. Nurs. **10**, 248–251 (2011)
12. Ekman, I., Busse, R., van Ginneken, E., Van Hoof, C., van Ittersum, L., Klink, A., Kremer, J.A., Miraldo, M., Olauson, A., De Raedt, W., Rosen-Zvi, M., Strammiello, V., Törnell, J., Swedberg, K.: Health-care improvements in a financially constrained environment. Lancet **387**, 646–647 (2016)
13. Fleming, F.: Evaluation methods for assessing value for money. Australas. Eval. Soc. (2013). http://betterevaluation.org/sites/default/files/Evaluating%20methods%20for%20asse ssing%20VfM
14. Gold, M.R., Siegel, J.E., Russell, L.B., Weinstein, M.C., Russell, L.B.: Cost-Effectiveness in Health and Medicine. Oxford University Press, Oxford (1996)
15. Gyllensten, H., Björkman, I., Jakobsson, E., Ekman, I., Jakobsson, S.: A national research centre for the evaluation and implementation of person-centred care: content from the first interventional studies. Health Expect. **23**, 1362–1375 (2020)
16. Hull, L.: Implementation Science Research Development (ImpRes) Tool: A Practical Guide to Using the ImpRes Tool. King's Improvement Science, London (2018)
17. ICAI: ICAI's Approach to Effectiveness and Value for Money. Independent Commission for Aid Impact (ICAI) (2011)
18. ICAI: DFID's approach to value for money in programme and portfolio management: a performance review (2018)
19. Lane-Fall, M.B., Curran, G.M., Beidas, R.S.: Scoping implementation science for the beginner: locating yourself on the "subway line" of translational research. BMC Med. Res. Methodol. **19**, 133 (2019)

20. Lavallee, D.C., Chenok, K.E., Love, R.M., Petersen, C., Holve, E., Segal, C.D., Franklin, P.D.: Incorporating patient-reported outcomes into health care to engage patients and enhance care. Health Aff. **35**(4), 575–582 (2016)
21. Lloyd, H.M., Ekman, I., Rogers, H.L., Raposo, V., Melo, P., Marinkovic, V.D., Buttigieg, S.C., Srulovici, E., Lewandowski, R.A., Britten, N.: Supporting innovative person-centred care in financially constrained environments: the WE CARE exploratory health laboratory evaluation strategy. Int. J. Environ. Res. Public Health **17**(9), 3050 (2020)
22. Lorenzoni, L., Murtin, F., Springare, L.-S., Auraaen, A., Daniel, F.: Which policies increase value for money in health care? (2018)
23. MacKinnon III, G.E.: Understanding Health Outcomes and Pharmacoeconomics. Jones & Bartlett Publishers, Burlington (2011)
24. Mounier, E.: Personalism. University of Notre Dame Press, Notre Dame (1952)
25. Nilsen, P.: Making sense of implementation theories, models and frameworks. Implement. Sci. **10**, 53 (2015)
26. OECD: Health Care Systems: Getting more Value for Money. Author Paris, France (2010)
27. Raleigh, V., Bardsley, M., Smith, P., Wistow, G., Wittenberg, R., Erens, B., Mays, N.: Integrated care and support Pioneers: indicators for measuring the quality of integrated care. Policy Innovation Research Unit, London School of Hygiene & Tropical Medicine, London (2014)
28. Ricoeur, P.: Oneself as another ((Soi-même comme un autre)). Trans. Kathleen Blamey. University of Chicago Press, Chicago 1992 (1990)
29. Santana, M.J., Manalili, K., Jolley, R.J., Zelinsky, S., Quan, H., Lu, M.: How to practice person-centred care: a conceptual framework. Health Expect. 1–12 (2017)
30. Santana, M.J., Ahmed, S., Lorenzetti, D., Jolley, R.J., Manalili, K., Zelinsky, S., Quan, H., Lu, M.: Measuring patient-centred system performance: a scoping review of patient-centred care quality indicators. BMJ Open **9**, e023596 (2019). https://doi.org/10.1136/bmjopen-2018-023596
31. Schlesinger, M., Grob, R., Shaller, D.: Using patient-reported information to improve clinical practice. Health Serv. Res. **50** (Suppl 2), 2116–2154 (2015). https://doi.org/10.1111/1475-6773.12420
32. Smith, P.C.: Measuring Value for Money in Healthcare: Concepts and Tools. The Health Foundation, London (2009)
33. Sorenson, C., Drummond, M., Kanavos, P.: Ensuring Value for Money in Health Care: The Role of Health Technology Assessment in the European Union. WHO Regional Office Europe (2008)
34. Tomaselli, G., Buttigieg, S.C., Rosano, A., Cassar, M., Grima, G.: Person-centered care from a relational ethics perspective for the delivery of high quality and safe healthcare: a scoping review. Front. Public Health **8**, 44 (2020)

Chapter 4
Person-Centred Care Interventions in Pharmaceutical Care

Valentina Marinkovic, Marina Odalovic, Ivana Tadic, Dusanka Krajnovic, Irina Mandic, and Heather L. Rogers

Abstract This chapter is divided into four sections. The first section introduces the concept of person-centred care within pharmaceutical care delivery and provides a historical context. The second section focuses on the professionals and explores the role of person-centred pharmaceutical care as part of multi-disciplinary health services delivery teams. The third section focuses on the patient and describes the role of health literacy in the implementation of person-centred pharmaceutical care. The last section examines E-pharmacy services and the implementation of telepharmacy with implications for person-centred care.

Keywords Pharmaceutical care · Person-centered care · Pharmaceutical intervention · Health professionals team · Patient · e-pharmacy

V. Marinkovic (✉) · M. Odalovic · I. Tadic · D. Krajnovic
Faculty of Pharmacy, Department of Social Pharmacy and Pharmaceutcal Legislation, University of Belgrade, Belgrade, Serbia
e-mail: valentina.marinkovic@pharmacy.bg.ac.rs

M. Odalovic
e-mail: marina.odalovic@pharmacy.bg.ac.rs

I. Tadic
e-mail: ivana.tadic@pharmacy.bg.ac.rs

D. Krajnovic
e-mail: dusica.krajnovic@pharmacy.bg.ac.rs

I. Mandic
Slovenian Pharmaceutical Society, Ljubljana, Slovenia

H. L. Rogers
BioCruces Health Research Institute, Bilbao, Spain

© The Author(s) 2022
D. Kriksciuniene and V. Sakalauskas (eds.), *Intelligent Systems for Sustainable Person-Centered Healthcare*, Intelligent Systems Reference Library 205,
https://doi.org/10.1007/978-3-030-79353-1_4

53

4.1 Introduction

Pharmaceutical care is a diverse concept. Over the last 50 years, many researchers and practitioners tried to define pharmaceutical care [1]. One of the first and widely cited is Hepler's and Strand's definition. It is cited, for example, in the recently adopted Resolution on the implementation of pharmaceutical care for the benefit of patients and health services [18], Resolution CM/Res 2020). According to that definition, pharmaceutical care is "the responsible provision of drug therapy for the purpose of achieving definite outcomes that improve a patient's quality of life" and it "involves the process through which a pharmacist co-operates with a patient and other professional in designing, implementing and monitoring a therapeutic plan that will produce specific therapeutic outcomes for the patient" [18]. The pharmacists' role in the pharmaceutical care provision has additionally been evaluated and described by Pharmaceutical Care Network Europe as "the pharmacist's contribution to the care of individuals in order to optimize medicines use and improve health outcomes" [1].

The main elements of pharmaceutical care involve the central role of the pharmacist and the patient-centered care approach, collaboration with carers, prescribers and other health care professionals (integrated care), prevention, detection and resolution of medication-related problems, and taking responsibility for optimising medication use in order to improve a patient's health outcomes and quality of life (Resolution CM/Res 2020). In order to address noted elements, the following activities are purposed within the process of pharmaceutical care delivery: (i) assessment of patient's medication needs and health status, (ii) identification and prioritisation of medication-related problems, (iii) selection of intervention(s) and formulation of pharmaceutical care plan, (iv) patient agreement, implementation and monitoring, and (v) follow-up. Proposed activities results in pharmaceutical care benefits described in the literature, involving revealing of patients' medication needs and drug related problems, optimising medication use and improve patients' quality of life.

4.2 Person Centred Care Model in Pharmaceutical Care

Some pharmacy personnel have begun to integrate person-centred care to enhance pharmaceutical care delivery. Key aspects of person-centred care within pharmaceutical care "include listening to the individual to understand their perspective, providing information in a manner which enables the person to make informed decisions and supporting them to develop goals relating to their lifestyle, health and medicines".

Barnet [3] described a model of person-centered care development and implementation within pharmacy practice in Great Britain. As a foundation for implementation of such model, two prerequisites were addressed: (i) the concept of self-care and self-management was introduced via regulations set out by the National Health System Plan in 2000, and (ii) the acceptance of the concept by leading authorities, such as the World Health Organisation [59]. The Royal Pharmaceutical Society (RPS) of Great Britain played an essential role in the implementation of person-centered care in the pharmaceutical sector. The RPS accepted and supported the concept within the publication "Now or never: shaping pharmacy for the future" [53], highlighting how pharmacists can help and maintain self-management. The RPS also issued the guideline "Medicines Optimisation: helping patients to make the most of medicines" where patient experience is listed as the first of four key principles of medicines optimisation [42]. Two years later, the RPS suggested measures to implement person-centered care delivery through pharmacies, primarily to patients with long-term conditions [3]. The concept was also incorporated in the publication "Standards for pharmacy professionals", defined by the General Pharmaceutical Council [14]. Further efforts were made by The Centre for Postgraduate Pharmacy Education to provide continuing professional education to support the transition to the active participation of people in their own care, shared decision-making between healthcare professionals and users of healthcare services and the introduction of health coaching to support medicines optimisation and improve adherence [3].

The capability of pharmacists to deliver person-centered care as a concept which integrates the person's needs, values and preferences is well-established within the main elements of pharmaceutical care. It is predominantly reflected through taking the responsibility for optimising medication use in order to improve health outcomes and quality of life. Accordingly, many pharmaceutical care services were suggested as convenient for delivery of person-centered care, such as medication reviews and helping people with discharge medicines when leave hospital, broader support for physical and mental wellbeing and inhaler technique support, etc.

However, several challenges in delivery of person-centered pharmaceutical care have been recognized. People are not familiar with pharmaceutical care services and support which can be accessed in community pharmacies. To address this, efforts should be made to raise the awareness of pharmacists and the services they offer. Collaboration with the voluntary and community sectors has been seen as the model for raising the awareness among people with long-term conditions about the support that pharmacists can offer. Non-governmental organisations, such as different associations of patient groups, can play an important role. Pharmacists need to be recognized as members of multidisciplinary teams, and collaboration with general practitioners must be improved (RPS 2017).

In summary, the contributions and the support of regulatory authorities and professional associations, community, other healthcare professionals and educational organisations are essential in the development, implementation and sustainability of person-centred pharmacy services. These elements represent pillars of person-centred pharmaceutical care, which are presented in the model illustrated in the Fig. 4.1.

There are promising indicators regarding the implementation of person-centred principles in pharmacy practice. Twigg and colleagues report on the Pharmacy Care Plan Service, introduced in Great Britain as a new community pharmacy intervention. It includes use of the Patient Activation Measure, a new tool for health professionals aimed to tailoring advice to individual needs of person. As part of the service,

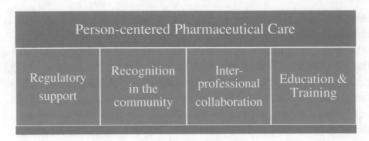

Fig. 4.1 Person-centered pharmacy service implementation and sustainability model

patients and pharmacists identify and agree on patient health goals. Patients have regular consultations with a pharmacist, who coaches and motivates the patients to enhance their quality of life through improved management of symptoms, lifestyle choices, weight loss and resulting improvement in health conditions. In conclusion, this example of implementation of person-centred pharmacy care in practice demonstrates that pharmacists can successfully recruit a large number of patients who are appropriate for such a service. Patients are willing to identify goals with the pharmacist. Future research will examine the impact of this service on health outcomes, but it is reasonable to argue that benefits will result from this service. If the majority of patient goals are met, they are likely to lead to improvements in quality of life for patients and their families.

4.3 Pharmacists as Collaborators in Multi-disciplinary Health Services Delivery Teams

The period between 2010 and 2020 will be remembered as a period of transformation of health care systems in their orientation to people-centered care (PCC) [63]. Providing services tailored to the identified people's needs has become a challenge for all healthcare professionals and health care systems. The integrated health care is a complex approach and requires consideration of persons' and population health needs, implementation of services related to identified needs, and transformation of health systems in order to provide customized services and strategic change management [61].

Integrated health services are defined, according to the WHO [61], as "an approach to strengthen people-centred health systems through the promotion of the comprehensive delivery of quality services across the life-course, designed according to the multi-dimensional needs of the population and the individual and delivered by a coordinated multi-disciplinary team of providers working across settings and levels of care". Integrated health services should be managed in the most cost-effective manner and should deliver optimal health outcomes [61]. According to Fulop's typology of integrated care, such "integration" might be established on four levels where the "service" integration level requires multi-disciplinary inter-sectoral and multi-sectoral teams of professionals to provide appropriate clinical services [28, 61]. Most definitions of pharmaceutical care focus on the improvement of patient health outcomes. Yet, the providers of pharmaceutical care vary from "anyone", to "pharmacists with their team", to "pharmacists only" to "practitioners", depending on the definition [1]. Ultimately, enhanced pharmacist-physician collaboration will allow both groups of professionals to provide effective pharmaceutical care that results in optimal health outcomes for the patient.

According to McDonough and Doucette [30], the collaboration between pharmacists and general practitioners can be described in five progressive stages: (1) "professional awareness", (2) "professional recognition", (3) "exploration and trial", (4) "expansion of professional relationships" and (5) "commitment to the collaborative working relations". The first stage represents the lowest degree of interaction, while, in the second stage, the collaboration is initiated from one party (mostly pharmacists) without any recognition of collaboration benefits by physicians. In the last stage, Stage 5, the pharmacist-physician collaboration is strong and there are mutual benefits for both parties with a high degree of trust and respect established. The model identifies other factors that can affect the pharmacist-physician collaboration, including individual personal characteristics, context characteristics (e.g., practices and settings), and social interaction (exchange) characteristics of collaborators [30].

Bradley and colleagues [7] propose an extended conceptual model that describes the levels of collaboration between pharmacists and general practitioners using a matrix. One side of the matrix indicates the elements deemed important for the collaboration, which include locality, service provision, trust, "knowing" each other, communication, professional roles and professional respect. The other side of the matrix describes the levels of collaboration for each element: isolation, communication and collaboration [7]. Rathbone and colleagues [45] examined collaboration between pharmacists and general practitioners with the specific aim to support patient medication adherence. They found various factors that influenced successful collaboration: proactive communication, direct communication (including electronic communication in situations where face-to-face communication is not possible), regular interactions, location, perceptions of credibility and common vision [45].

Some studies investigated physician and pharmacist perceptions about barriers to collaboration to achieve the best patient health outcomes. In one study, general practitioners identified the two most important activities of pharmacists as dispensing medicines and helping patients to achieve improved medication adherence. The two most important barriers for pharmacist-physician collaboration were a perceived lack of need to collaborate with a variety of health care professionals and a lack of compensation for collaboration [23]. Another study found that the image of a pharmacist as a "shopkeeper" influences general practitioners' opinions about extending pharmacy services or prescribing rights. Pharmacists viewed general practitioners' secretaries as barriers to establishing collaboration. The "lack of awareness" between pharmacists and general practitioners was stressed as another barrier. General practitioners expressed that they did not know what skills and training pharmacists have, and pharmacists felt that general practitioners underestimated them and their contributions to patients' health care [19].

The Kaiser Permanente (USA) population-based model is primarily based on multi-disciplinary medical practice where pharmacists are active members of professional health care teams [17]. The Kaiser Permanente model is oriented towards optimal patient care via multi-disciplinary practice. As such, this model does not include a strict division of services that should be provided at the primary, secondary or tertiary levels. Provision of services from mixed health care levels led to proven cost-effective services. The model is based on population and case management. The population management aspect is focused on services provided to patients depending on the complexity of their health condition and risk factors. Therefore, the *population management* aspect includes services within health promotion and disease prevention, self-management support and disease management. These services apply to the majority of the population. The *case management* aspect includes services provided to a sub-set of the population. This is the smallest part of all patients and includes services for treatment of severe complications [62]. Applying these principles resulted in fewer hospitalized patients comparing to United Kingdom National Health System. The role of pharmacists in such a model is of great importance, as they can make important contributions in both population and case management levels [12, 29, 63].

The collaborative practice of pharmacists and medical practitioners is widely recognized in the care of patients with chronic conditions such as diabetes, hypertension, asthma, and chronic obstructive pulmonary disease. Successful collaboration may lead to better care of patients, improved disease management and, ultimately, lower health care costs [20, 31, 39, 50]. Nowadays, in many countries, medications can be prescribed via electronic prescriptions and can be dispensed repeatedly for a period of several months without need for a visit with a general practitioner. In these cases, the role of the pharmacist is crucial in monitoring of health outcomes, medication optimization, managing drug-related problems and patient education [6].

In recent years organizations such as the World Health Organization (WHO) and the Organisation for Economic Co-operation and Development (OECD) recognized community pharmacists as health care professionals who can provide direct health care services of high quality to patients by themselves or in collaboration with other health care professionals [54, 65]. Although the primary role of community pharmacists is supply of medicines, which is very important for access to and safety of medicines, there are many other services that pharmacists can provide. In fact, the Pharmaceutical Group of the European Union (PGEU) has identified 38 different pharmacy services in PGEU member countries. In line with the Kaiser Permanente model principles, all services were stratified as dispensing and related services, services in health promotion and disease prevention, screening and referral services, and disease and individual case management services [21]. The effectiveness of many of these services could be enhanced if they were implemented in collaboration with other health care professionals.

The pandemic of coronavirus disease revealed that European health systems are also facing a shortage of health care professionals [34]. This trend is expected to increase [60, 62]. At the same time, the number of chronic patients is increasing every year [55]. Pharmacists possess specific knowledge on medicines and patient care, and they develop competencies throughout their careers to be able to provide effective and efficient pharmaceutical care. Since pharmaceutical and medicine are related sciences, there is a partial overlap in a number of relevant services delivered to patients with chronic diseases. In conclusion, it is of paramount importance that pharmacists begin to be recognized as health professionals who can provide more services and work in teams with other health care professionals. Within the PGEU vison for community pharmacy in Europe, patient-centred care and person-centred care provided in multi-disciplinary collaborative care teams of pharmacists and other health care professionals on different levels and in different health care settings are recognized as a challenge and an opportunity to foster quality of care and patient safety (Pharmaceutical Group of European Union).

4.4 Patients as Collaborators in Their Own Care: The Role of Health Literacy and Pharmacotherapy Literacy

How person-centred care (PCC) is defined, applied and measured differs within and across countries. There is general agreement that PCC refers to 'care that is centred on the person' or care of the 'whole person' [10, 24, 33]. A substantial international body of work describes person-centred healthcare as a multi-dimensional concept that delivers care responsive to people's individual abilities, preferences, lifestyles and goals [33, 35, 49]. According to *The Health Foundation* (2014) person-centred care could mean, at the very least, four different things: (i) affording people dignity, respect and compassion, (ii) offering coordinated care, support or treatment, (iii) offering personalised care, support or treatment and (iv) being enabling [11]. Achieving person-centred healthcare requires a system that supports people in making informed decisions about and successfully managing their own health and care, including therapy choices and choosing when to let others act on their behalf. The patient experience of receiving care within the spectrum of healthcare is core to the concept of person-centred care.

In order for patients to play an important role in their own health care, as advocated in person-centred care, health literacy is a key factor influencing this capacity. Health literacy has been recognized as an important component of health care, and the World Health Organisation describes health literacy as "the cognitive and social skills which determine the motivation and ability of individuals to gain access to, understand and use information in ways which promote and maintain good health" [26]. Health literacy is a multi-dimensional concept composed of a variety of cognitive, affective, social, and personal skills and attributes. Buchbinder and colleagues [8] identified seven key abilities required for an individual to be able to seek, understand, and use health information: (1) knowing when to seek health information,(2) knowing where to seek health information,(3) verbal communication skills; (4) assertiveness; (5) literacy skills; (6) capacity to process and retain information; and (7) skills in applying health-related information.

Information about the health literacy of people in a community can offer better insight into the challenges people experience when trying to access and engage with healthcare services. Low literate patients may also provide poor self-assessment of their health, wellbeing or engagement with their care. Individuals with limited health literacy experience difficulties in understanding medicines labels, which may include misunderstanding of written medicine instructions, inadequate adherence to prescribed regimens, and inability to follow advice from health professionals regarding side effects and possible contraindications [25, 57]. Studies suggest that differences in health literacy abilities may explain observed health inequalities among people of different race and with different levels of educational levels [5, 37, 52, 58, 64]. Therefore, developing interventions to address low health literacy offers an opportunity to improve person-centred healthcare, health outcomes and reduce health inequalities. There are many measures to assess health literacy at the individual (clinical) and population level. The most widely used clinical measures include the Rapid Estimate of Adult Literacy in Medicine (REALM), which tests ability to read and pronounce a list of words, and the Test of Functional Health Literacy in Adults (TOFHLA), which tests reading comprehension and numeracy, whereas the Newest Vital Sign (NVS) is designed to be a quick clinical screening instrument.

One of the most comprehensive measures of health literacy at the population level relevant to person-centered healthcare is the Health Literacy Questionnaire (HLQ) developed by Deakin University and Monash University in Australia [4, 16]. It consists of 44 questions within nine domains: 1. Feeling understood and supported by healthcare providers 2. Having sufficient information to manage my health 3. Actively managing my health 4. Social support for health 5. Appraisal of health information 6. Ability to actively engage with healthcare providers 7. Navigating the healthcare system 8. Ability to find good health information 9. Understanding health information well enough to know what to do. Osborne and colleagues [41] found it acceptable to patients and healthcare workers, but it has been designed and tested for use at the population level, not for use with individual patients. It was designed for self-administration using pen and paper and can also be interviewer-administered. Data derived from HLQ included both subjective (e.g. 'did you feel respected?') and objective (e.g. 'were you offered a care plan?') items, as well as general (e.g. 'how satisfied were you with your care?') and specific (e.g. 'how satisfied were you with family visiting arrangements?') items [41].

Medication or pharmacotherapy literacy is a sub-area of health literacy. It is partic-ularly important to patient safety in the context of person-centred care. Many studies have stressed that inadequate levels of medication literacy may negatively affect phar-macotherapy outcomes and safety of care delivery, particularly in areas of greater social deprivation and where significant health inequalities exist. Any analysis of pharmacotherapy literacy must navigate a quite heterogeneous body of literature and commentary and innovative practice. "Medication literacy" was generally defined as the individual's ability to understand and act on medication-related information. Pouliot and colleagues [44] defined it as "the degree to individuals can obtain, comprehend, communicate, calculate and process patient specific information about their medications, and make informed medication and health decisions in order to safely and effectively use their medications- regardless of the mode by which the content is delivered " [44]. The overlap with person-centred care is highlighted in this definition of pharmacotherapy literacy as "an individual's capacity to obtain, evaluate, calculate, and comprehend basic information about pharmacotherapy and pharmacy related services necessary to make appropriate medication-related decisions, regardless of the mode of content delivery (e.g., written, oral, visual images and symbols)" [25, 57]. Although medication literacy is a relatively recent concept, several tools have already been developed to assess it [57]. In conclusion, the linkage between pharmacotherapy literacy and person-centred care approaches is currently relatively under-developed and may be an important area for development in the future [25, 43, 44, 46, 57].

4.5 E-Pharmaceutical Service and Implications for Person-Centred Care

Many health systems in Europe are characterized by a shortage of health workers, overwork and high costs, all of which lead to unsustainability in the long run [38]. Population aging is a long-term trend that began in Europe several decades ago. The age structure of the population has been reshaped, so there is an increasing percentage of the elderly and consequently a decreasing percentage of working people in the total population [13]. On the other hand, beneficiaries of health care services are insufficiently informed or uninformed, they usually wait too long for specialist examinations, interventions and have the feeling that doctors do not pay enough attention to them. This current status couldn't achieve Person-centred care framework.

Might telemedicine/telepharmacy be one solution to address some of these issues identified by health system professionals and users?

Historically, the beginnings of telemedicine can be traced back to the nineteenth century and the invention of the electric telegraph and telephone, which allowed doctors and patients to communicate remotely [22]. For the first time, in 1959, the University of Nebraska used two-way (interactive) television to transmit neurological examinations to students [51]. The American Telemedicine Association (ATA) was founded in 1993 with the aim of promoting access to telemedicine care through telecommunications technology [27]. In the early twenty-first century a wider use of the Internet made the development of telemedicine and telepharmacy possible, together with advancements in their regulations and standards. Despite the legal basis in European Union regulations, where telemedicine falls under health and information services, international and national legal frameworks for the use of telemedicine are lacking. To overcome these challenges, telemedicine must be regulated by comprehensive legal guidelines which must be placed within a single international legal framework [47].

Operatively, telemedicine covers two broad areas. The first is the virtual interaction between the patient and the health care provider, and the second is the flow of information [36].

Telepharmacy, which is a part of telemedicine, may help improve pharmacy service coverage regardless of the pharmacist's location. Telepharmacy considers activities such as: electronic data entry, prescription order verification, centrally, online benefit adjudication, medication dispensing, and telehealth consultation with medication use evaluation using computer technology [48].

One of the first studies demonstrating the person-centered nature of pharmaceutical care, delivered via a telepharmacy service model, was introduced by Sankaranarayanan and colleagues [48].

Regarding this telepharmacy service model, three definitions at the intervention level have been introduced: the patient-centered medication management level, the health system-centered medication use process level and the remote pharmacist role across hospitals with and without an on-site hospital pharmacist.

An interprofessional and collaborative practice between physician and pharmacist is one of the perquisites of person-centered care [9]. Using telemedicine/ telepharmacy, communication is facilitated and incredibly useful for improving patients' health. The inclusion of telepharmacy may expand the reach of the pharmacist's intervention and provide pharmacy operations and patient care at a distance with further benefits for patients and their managing physicians [40].

An e-pharmaceutical intervention improved health outcome in many clinical areas: anticoagulant therapy monitoring, oncology, diabetes, hypertension [15]. E-pharmaceutical interventions was an efficient, safe, and cost-effective method for implementing person-centered approach.

The global Covid-19 virus pandemic demonstrated the urgent need for implementation of telemedicine services. In the Republic of Slovenia, for example, new providers of telemedicine services grew each day of the pandemic, while the Public Institute Pharmacy Ljubljana has only recently introduced telepharmacy services [32].

Recent research on the attitudes of Slovenian pharmacists indicate that telepharmacy services should be charged for (Mandic 2020, unpublished data). In other words, they suggested a model of financing separate from compulsory health insurance in to ensure the financial sustainability of the new service and allow it to always be available to patients. Although there are important challenges with regards to the systemic, legal regulation of this area and the protection of personal data, as discussed earlier in the section, the study participants endorsed the need to introduce new services into the pharmacy system and, in general, a very positive attitude towards the introduction of telepharmaceutical services. Their motivation for learning and acquiring new skills was clear, and readiness for change is an important precondition for the implementation of new services. They see telepharmacy as a potential for progress and further development and contribution to the health system in the Republic of Slovenia, as well.

This study of Slovenian pharmacists also provided positive results regarding digital literacy. Pharmacists indicated that they actively use digital communication channels and were confident in their levels of digital literacy. If educating elderly patients were possible or telepharmaceutical services were set up so that they are extremely easy to use, the pharmacists who participated in the survey did not see any major obstacles in providing new services based on digital communication channels. They expressed a clear interest in using digital technologies for the purpose of providing a service that would be more accessible to patients, but also to enable greater and easier interaction to pharmacists, allowing greater supervision and control over the health of their patients. Personal contact is certainly best for conversation, counseling and understanding the needs of patients. However, if personal contact is hindered for some reason, the pharmacists believed that they could make up for this by offering a slightly higher degree of empathy because the patients were already comfortably at home. Telepharmaceutical services should not additionally burden pharmacists nor require too much administration, and certainly those services should be financially valued in some way (Mandic 2020, unpublished data).

In order to fully develop and implement telepharmacy services with a health system, in context of person-centered care, certain resources are required. For instance, adequate standards, education and training of pharmacists are needed [2].

Telepharmacy, i.e. e-pharmacy, can take place through special platforms (applications or sites), but this is not a necessary prerequisite. Implementation can start at a simpler level, using already available resources, such as ordinary telephone conversations or calls via free online communication channels (Skype, Viber, Whatsapp, Zoom, Webex, etc.).

Any new service, including telepharmacy, should be simple, safe, financially viable and, above all, assist both patients and pharmacists in achieving their joint goal: person in the center of the healthcare system.

4.6 Conclusion

This chapter examined person-centred care within pharmaceutical care delivery in detail. The importance of the pharmacist as part of a multi-disciplinary care team and the need for close collaboration with general practitioners, including barriers to such collaboration, was examined. Then the role of the patient in person-centred pharmaceutical care, with an emphasis on health literacy, and medication literacy in particular, was explored. The chapter concluded with an overview of telemedicine and e-pharmacy or telepharmacy services, with implications for person-centred care.

Acknowledgements This publication is based upon work from COST Action "**European Network for cost containment and improved quality of health care-CostCares**" (CA15222), supported by COST (European Cooperation in Science and Technology)

COST (European Cooperation in Science and Technology) is a funding agency for research and innovation networks. Our Actions help connect research initiatives across Europe and enable scientists to grow their ideas by sharing them with their peers. This boosts their research, career and innovation.

https://www.cost.eu

Funded by the Horizon 2020 Framework Programme of the European Union

References

1. Allemann, S.S., van Mil, J.W., Botermann, L., Berger, K., Griese, N., Hersberger, K.E.: Pharmaceutical care: the PCNE definition 2013. Int. J. Clin. Pharm. **36**(3), 544–555 (2014). https://doi.org/10.1007/s11096-014-9933-x
2. Baldoni, S., Amenta, F., Ricci, G.: Telepharmacy services: present status and future perspectives: a review. Medicina (Kaunas) **55**(7), 327 (2019). https://doi.org/10.3390/medicina55070327
3. Barnet N.: Person-centred over patient-centred care: not just semantics. How does the technology giant's marketing approach relate to person-centred care in health? Pharm. J. (2018)
4. Batterham, R.W., Buchbinder, R., Beauchamp, A., Dodson, S., Elsworth, G.R., Osborne, R.H.: The OPtimising HEalth LIterAcy (Ophelia) process: study protocol for using health literacy profiling and community engagement to create and implement health reform. BMC Public Health **14**, 694 (2014). https://doi.org/10.1186/1471-2458-14-694
5. Berkman, N.D., Sheridan, S.L., Donahue, K.E., Halpern, D.J., Crotty, K.: Low health literacy and health outcomes: an updated systematic review. Ann. Intern. Med. **155**(2), 97–107 (2011). https://doi.org/10.7326/0003-4819-155-2-201107190-00005
6. Bond, C., Matheson, C., Williams, S., Williams, P., Donnan, P.: Repeat prescribing: a role for community pharmacists in controlling and monitoring repeat prescriptions. Br. J. Gen. Pract. **50**(453), 271–275 (2000)
7. Bradley, F., Ashcroft, D.M., Noyce, P.R.: Integration and differentiation: a conceptual model of general practitioner and community pharmacist collaboration. Res. Soc. Adm. Pharm. **8**(1), 36–46 (2012). https://doi.org/10.1016/j.sapharm.2010.12.005
8. Buchbinder, R., Batterham, R., Ciciriello, S., Newman, S., Horgan, B., Ueffing, E., Rader, T., Tugwell, P.S., Osborne, R.H.: Health literacy: what is it and why is it important to measure? J. Rheumatol. **38**(8), 1791–1797 (2011). https://doi.org/10.3899/jrheum.110406
9. Bugnon, O., Hugentobler-Hampaï, D., Berger, J., Schneider, M.P.: New roles for community pharmacists in modern health care systems: a challenge for pharmacy education and research. CHIMIA (Aarau) **66**(5), 304–307 (2012). https://doi.org/10.2533/chimia.2012.304
10. Care Quality Commission: The State of Healthcare and Adult Social Care in England - An Overview of Key Themes in Care 2009/10. Care Quality Commission, London (2010)
11. Collins, A.: Measuring What Really Matters. The Health Foundation, London (2014)
12. Curry, N., Ham, C.: Clinical and service integration. The route to improved outcomes (2010)
13. Eurostat Statistics Explained. Available at: https://ec.europa.eu/eurostat/statistics-explained/index.php?title=Population_structure_and_ageing/sl#Pretekli_in_prihodnji_trendi_staranja_prebivalstva_v_EU. Accessed 20 Dec 2020
14. General Pharmaceutical Council: Standards for pharmacy professionals (2017). Available at: https://www.pharmacyregulation.org/sites/default/files/standards_for_pharmacy_professionals_may_2017_0.pdf. Accessed 20 Dec 2020
15. Hawes, E.M., Lambert, E., Reid, A., Tong, G., Gwynne, M.: Implementation and evaluation of a pharmacist-led electronic visit program for diabetes and anticoagulation care in a patient-centered medical home. Am. J. Health Syst. Pharm. **75**(12), 901–910 (2018). https://doi.org/10.2146/ajhp170174

16. Hawkins, M., Gill, S.D., Batterham, R., Elsworth, G.R., Osborne, R.H.: The health literacy questionnaire (HLQ) at the patient-clinician interface: a qualitative study of what patients and clinicians mean by their HLQ scores. BMC Health Serv. Res. **17**(1), 309 (2017). https://doi.org/10.1186/s12913-017-2254-8

17. Helling, D.K., Nelson, K.M., Ramirez, J.E., Humphries, T.L.: Kaiser Permanente Colorado region pharmacy department: innovative leader in pharmacy practice (2003). J. Am. Pharm. Assoc. **46**(1), 67–76 (2006). https://doi.org/10.1331/154434506775268580

18. Hepler, C.D., Strand, L.M.: Opportunities and responsibilities in pharmaceutical care. Am. J. Hosp. Pharm. **47**(3), 533–543 (1990)

19. Hughes, C.M., McCann, S.: Perceived interprofessional barriers between community pharmacists and general practitioners: a qualitative assessment. Br. J. Gen. Pract. **53**(493), 600–606 (2003)

20. Hwang, A.Y., Gums, T.H., Gums, J.G.: The benefits of physician-pharmacist collaboration. J. Fam. Pract. **66**(12), E1–E8 (2017)

21. Institute for Evidence Based Health: Pharmacy services in Europe: evaluating trends and value report. Lisboa (2020)

22. Institute of Medicine (US): Committee on evaluating clinical applications of telemedicine. In: Field, M.J. (eds.) Telemedicine: A Guide to Assessing Telecommunications in Health Care. National Academies Press (US), Washington (DC) (1996). Available at: https://www.ncbi.nlm.nih.gov/books/NBK45445. Accessed 20 Dec 2020

23. Kelly, D.V., Bishop, L., Young, S., Hawboldt, J., Phillips, L., Keough, T.M.: Pharmacist and physician views on collaborative practice: findings from the community pharmaceutical care project. Can. Pharm. J. (Ott) **146**(4), 218–226 (2013). https://doi.org/10.1177/171516351349 2642

24. Kidd M.: Personal correspondence. World Organisation of Family Doctors (WONCA) and Flinders University, Australia, July 2015

25. King, S.R., McCaffrey, D.J., Bouldin, A.S.: Health literacy in the pharmacy setting: defining pharmacotherapy literacy. Pharm. Pract. **9**(4), 213–220 (2011). https://doi.org/10.4321/s1886-36552011000400006

26. Krajnović, D., Ubavić, S., Bogavac-Stanojevic, N.: Pharmacotherapy literacy and parental practice in use of over-the-counter pediatric medicines. Medicina (Kaunas) **55**(3), 80 (2019). https://doi.org/10.3390/medicina55030080

27. Krupinski, E.A., Antoniotti, N., Bernard, J.: Utilization of the American telemedicine association's clinical practice guidelines. Telemed. E-Health **19**(11), 846–851 (2013). https://doi.org/10.1089/tmj.2013.0027

28. Lewis, R., Rosen, R., Goodwin, N., Dixon, J.: Where next for integrated care organizations in the English NHS? The Nuffield Trust, London (2010)

29. Light, D., Dixon, M.: Making the NHS more like Kaiser Permanente. BMJ **328**(7442), 763–765 (2004). https://doi.org/10.1136/bmj.328.7442.763

30. McDonough, R., Doucette, W.: Developing collaborative working relationships between pharmacists and physicians. J. Am. Pharm. Assoc. **41**(5), 682–692 (2001)

31. Mes, M.A., Katzer, C.B., Chan, A.H.Y., Wileman, V., Taylor, S.J.C., Horne, R.: Pharmacists and medication adherence in asthma: a systematic review and meta-analysis. Eur. Respir. J. **52**(2), 1800485 (2018). https://doi.org/10.1183/13993003.00485-2018

32. Mestna občina Ljubljana <Ali ste vedeli:Lekarna Ljubljana prva v Sloveniji uvaja telefarmacevtsko svetovanje>. Available at: www.ljubljana.si/sl/moja-ljubljana/ali-ste-vedeli/ali-ste-ved eli-lekarna-ljubljana-prva-v-sloveniji-uvaja-telefarmacevtsko-svetovanje/. Accessed Oct 2020

33. Mezzich, J.E., Appleyard, J., Bothol, M., Ghebrehiwot, T., Groves, J., Salloum, I., van Dulmen, S.: Ethics in person centered medicine: conceptual place and ongoing developments. J. Pers.-Cent. Med. **3**(4), 255–257 (2013)

34. Michel, J.P., Ecarnot, F.: The shortage of skilled workers in Europe: its impact on geriatric medicine. Eur. Geriatr. Med. **11**(3), 345–347 (2020). https://doi.org/10.1007/s41999-020-003 23-0

35. Miles, A., Asbridge, J.E.: Clarifying the concepts, epistemology and lexicon of person-centeredness: an essential pre-requisite for the effective operationalization of PCH within modern healthcare systems. Eur. J. Pers. Cent. Healthc. **2**(1), 1–15 (2014). https://doi.org/10.5750/ejpch.v2i1.857
36. Nittari, G., Khuman, R., Baldoni, S., Pallotta, G., Battineni, G., Sirignano, A., Amenta, F., Ricci, G.: Telemedicine practice: review of the current ethical and legal challenges. Telemed. E-Health **26**(12), 1427–1427 (2020). https://doi.org/10.1089/tmj.2019.0158
37. Nutbeam, D.: The evolving concept of health literacy. Soc. Sci. Med. **67**(12), 2072–2078 (2008). https://doi.org/10.1016/j.socscimed.2008.09.050
38. OECD: Health workforce (2020). Available at: www.oecd.org/els/health-systems/workforce.htm. Accessed 20 Dec 2020
39. Omboni, S., Caserini, M.: Effectiveness of pharmacist's intervention in the management of cardiovascular diseases. Open Hear. **5**(1), e000687 (2018). https://doi.org/10.1136/openhrt-2017-000687
40. Omboni, S., Tenti, M., Coronetti, C.: Physician-pharmacist collaborative practice and telehealth may transform hypertension management. J. Hum. Hypertens. **33**(3), 177–187 (2019). https://doi.org/10.1038/s41371-018-0147-x
41. Osborne, R.H., Batterham, R.W., Elsworth, G.R., Hawkins, M., Buchbinder, R.: The grounded psychometric development and initial validation of the health literacy questionnaire (HLQ). BMC Public Health **13**, 658 (2013). https://doi.org/10.1186/1471-2458-13-658
42. Picton, C., Wright, H.: Medicines optimisation: helping patients to make the most of medicines. Royal Pharmaceutical Society (2013). Available at: www.rpharms.com/Portals/0/RPS%20document%20library/Open%20access/Policy/helping-patients-make-the-most-of-their-medicines.pdf. Accessed 20 Dec 2020
43. Pouliot, A., Vaillancourt, R.: Medication literacy: why pharmacists should pay attention. Can. J. Hosp. Pharm. **69**(4), 335–336 (2016). https://doi.org/10.4212/cjhp.v69i4.1576
44. Pouliot, A., Vaillancourt, R., Stacey, D., Suter, P.: Defining and identifying concepts of medication literacy: an international perspective. Res. Soc. Adm. Pharm. **14**(9), 797–804 (2018). https://doi.org/10.1016/j.sapharm.2017.11.005
45. Rathbone, A.P., Mansoor, S.M., Krass, I., Hamrosi, K., Aslani, P.: Qualitative study to conceptualise a model of interprofessional collaboration between pharmacists and general practitioners to support patients' adherence to medication. BMJ Open **6**(3), e010488 (2016). https://doi.org/10.1136/bmjopen-2015-010488
46. Raynor, D.K.: Addressing medication literacy: a pharmacy practice priority. Int. J. Pharm. Pract. **17**(5), 257–259 (2009)
47. Ryu, S.: Telemedicine: opportunities and developments in Member States: report on the second global survey on eHealth 2009 (Global Observatory for eHealth, Volume 2). Healthc. Inform. Res. **18**(2), 153–155 (2012). https://doi.org/10.4258/hir.2012.18.2.153
48. Sankaranarayanan, J., Murante, L., Moffett, L.: A retrospective evaluation of remote pharmacist interventions in a telepharmacy service model using a conceptual framework. Telemed. E-Health **20**(10), 893–901 (2014). https://doi.org/10.1089/tmj.2013.0362
49. Scholl, I., Zill, J.M., Härter, M., Dirmaier, J.: An integrative model of patient-centeredness– a systematic review and concept analysis. PLoS One **9**(9), e107828 (2014). https://doi.org/10.1371/journal.pone.0107828
50. Siaw, M.Y.L., Ko, Y., Malone, D.C., Tsou, K.Y.K., Lew, Y.J., Foo, D., Tan, E., Chan, S.C., Chia, A., Sinaram, S.S., Goh, K.C., Lee, J.Y.: Impact of pharmacist-involved collaborative care on the clinical, humanistic and cost outcomes of high-risk patients with type 2 diabetes (IMPACT): a randomized controlled trial. J. Clin. Pharm. Ther. **42**(4), 475–482 (2017). https://doi.org/10.1111/jcpt.12536
51. Simson, A.: A brief history of NASA's contributions to telemedicine (2013). Available at: www.nasa.gov/content/a-brief-history-of-nasa-s-contributions-to-telemedicine. Accessed 20 Dec 2020
52. Sørensen, K., Pelika, J.M., Röthlin, F., Ganahl, K., Slonska, Z., Doyle, G., Fullam, J., Kondilis, B., Agrafiotis, D., Uiters, E., Falcon, M., Mensing, M., Tchamov, K., van den Broucke, S.,

Brand, H., HLS-EU Consortium: Health literacy in Europe: comparative results of the European health literacy survey (HLS-EU). Eur. J. Public Health **25**(6), 1053–1058 (2015). https://doi.org/10.1093/eurpub/ckv043

53. Smith, J., Picton, C., Dayan, M.: Now or never: shaping pharmacy for the future. Royal Pharmaceutical Society (2013). Available at: www.rpharms.com/Portals/0/RPS%20document%20library/Open%20access/Publications/Now%20or%20Never%20-%20Report.pdf. Accessed 20 Dec 2020

54. The Organisation for Economic Co-operation and Development: Health at a Glance 2019, OECD Indicators. Pharmacists and Pharmacies (2019)

55. The Organisation for Economic Co-operation and Development: Health at a Glance 2019: OECD Indicators (2019)

56. The Royal Pharmaceutical Society: The role of pharmacy in delivering person-centred care. Available at: www.rpharms.com/resources/reports/role-of-pharmacy-in-delivering-person-centred-care. Accessed 20 Dec 2020

57. Ubavić, S., Bogavac-Stanojević, N., Jović-Vraneš, A., Krajnović, D.: Understanding of information about medicines use among parents of pre-school children in Serbia: parental pharmacotherapy literacy questionnaire (PTHL-SR). Int. J. Environ. Res. Public Health **15**(5), 977 (2018). https://doi.org/10.3390/ijerph15050977

58. United Nations Economic and Social Council (ECOSOC): Health literacy and the millennium development goals. United Nations Economic and Social Council (ECOSOC) regional meeting background paper (abstracted). J. Health Commun. **15**(Suppl 2), 211–223 (2010). https://doi.org/10.1080/10810730.2010.499996

59. World Health Organization: People-centred health care: a policy framework (2007). Available at: www.wpro.who.int/health_services/people_at_the_centre_of_care/documents/ENG-PCIPolicyFramework.pdf. Accessed 12 Dec 2020

60. World Health Organization: A universal truth: no health without a workforce (2014). Available at https://www.who.int/workforcealliance/knowledge/resources/hrhreport2013/en/. Accessed 12 Dec 2020

61. World Health Organization: The European framework for action on integrated health services delivery: an overview (2016). Available at https://www.euro.who.int/en/health-topics/Health-systems/health-services-delivery/publications/2016/the-european-framework-for-action-on-integrated-health-services-delivery-an-overview-2016. Accessed 12 Dec 2020

62. World Health Organization: Global strategy on human resources for health: workforce 2030 (2016). Available at https://www.who.int/hrh/resources/globstrathrh-2030/en/. Accessed 12 Dec 2020

63. World Health Organization: Integrated care models: an overview (2016). Available at https://www.euro.who.int/en/health-topics/Health-systems/health-services-delivery/publications/2016/integrated-care-models-an-overview-2016. Accessed 20 Dec 2020

64. World Health Organization: Shanghai declaration on promoting health in the 2030 agenda for sustainable development. Health Promot. Int. **32**(1), 7–8 (2017). https://doi.org/10.1093/heapro/daw10

65. World Health Organization: The legal and regulatory framework for community pharmacies in the WHO European region (2019). Available at https://apps.who.int/iris/handle/10665/326394. Accessed 20 Dec 2020

Part II
Efficiency Evaluation, Decision-Making and Sustainability in Person-Centred Healthcare

Chapter 5
Shared Decision Making

Valentina Marinkovic, Heather L. Rogers, Roman Andrzej Lewandowski,
and Ivana Stevic

Abstract This chapter is divided into three sections. The first section introduces the concept and models of shared decision-making as a framework of person-centered care. The second section focuses on multicriteria decision-making techniques in healthcare settings and literature review about multicriteria decision making analysis methods used in healthcare is presented. The third section introduces the ethical and practical considerations about shared decision-making in person-centered care. In this section, the patient narratives are included, as well as the barriers to implementation.

Keywords Shared decision-making · Person-centered care · Healthcare · Multicriteria decision-making

5.1 Introduction

Decision making (DM) is one of the most important activities in the healthcare system and medical practice. Because health outcomes are probabilistic, most decisions are made under conditions of uncertainty [30].

V. Marinkovic (✉) · I. Stevic
Department of Social Pharmacy and Pharmaceutcal Legislation, Faculty of Pharmacy, University of Belgrade, Belgrade, Serbia
e-mail: valentina.marinkovic@pharmacy.bg.ac.rs

I. Stevic
e-mail: ivana.stevic@pharmacy.bg.ac.rs

H. L. Rogers
BioCruces Health Research Institute, Bilbao, Spain

R. A. Lewandowski
Institute of Management and Quality Science, Faculty of Economics, University of Warmia and Mazury, Olsztyn, Poland
e-mail: rlewando@wp.pl

D. Kriksciuniene and V. Sakalauskas (eds.), *Intelligent Systems for Sustainable Person-Centered Healthcare*, Intelligent Systems Reference Library 205,
https://doi.org/10.1007/978-3-030-79353-1_5

Person-centred care is a valuable approach to improve health care outcomes, so involvement of the patient/person in health care decisions could be beneficial for all interested parties.

5.2 Shared Decision Making Models as a Framework of Person-Centered Care

Shared decision-making (SDM) can be analysed as a model of collaborative practice in which decision-making is delegated, shared and intertwined in all directions of the traditional value chain. At the primary level of health care, SDM models have a particularly difficult and demanding path from development to implementation, bearing in mind that they imply breaking the traditional monopolistic hierarchy of decision-making in which doctors' opinion was primary and almost predominant in decision making. Shared decision-making models, however, do not aim to degrade the role of any participant in the decision-making chain, or to strengthen another participant, but to increase the involvement of all participants in the decision-making chain, at all levels of decision-making.

Laws and professional guides have adopted SDM vocabulary: the World Health Organization considers "autonomy while respecting the involvement of individuals (patients) in their health choices" [31]. The guides from the health ministries of Canada, the United States, the United Kingdom and Australia describe and recommend SDM as part of health studies and vocational training programs, and an integral part of good health practice. For example, in the UK, the General Medical Council emphasizes: In whatever context health care decisions are made, it is necessary to work with patients in partnership to ensure a high level of health care while improving health outcomes. Finally, it is necessary to:

- Listen to patients and respect their views on their own health
- Talk to patients about their diagnosis, prognosis, therapy, and health care
- Share with patients the necessary information to be able to make decisions
- Maximize patients' opportunities as well as their ability to make decisions for themselves
- Respect patient decisions.

In 2008, the General Medical Council also said that non-compliance and absence of patient adherence could put the entire health system in danger.

As the biggest problem for the successful implementation of SDM, the American Health Association finds in the professional education of medical doctors, which teaches them that they must always have the right answer, as well as that they must always have the final decision, and operate separately from other decision-makers in the health system. Although significant progress has been made in interprofessional education and communication to this end, SDM is still underdeveloped despite several simulations and virtual patients as an approach to studying SDM [22].

Shared decision-making (SDM) has often been described in the context of various kinds of physician-patient relationships. One seminal paper on this topic [21] describes four types of models and their relationship to decision making regarding patient care:

(1)   In the paternalistic model of care, the physician is the patient's guardian [7, 59]. He/she determines the best course of action for a particular patient and presents information that will encourage the patient to consent. In extreme versions of this model, the physician takes an authoritative role and makes a decision for the patient, with the patient informed of the next steps.

(2)   In the informative model of care, the physician is a technical expert who informs and implements the patient's wishes [7, 59]. He/she communicates facts to the patient about the disease and various treatment options. All information related to the advantages and disadvantages of these options are presented to the patient who makes a decision, in accordance with his/her values, as to how to proceed.

(3)   In the interpretative model of care, the physician is a counsellor [40, 50]. In addition to providing information, he/she assists the patient in clarifying goals and values, and helps the patient to understand which treatment options might align with these aims. Through this joint process, the patient learns more about himself/herself and makes a decision. In extreme versions of this model, the physician looks at the patient's life as a narrative whole, and then identifies the patient's values and priorities.

(4)   In the deliberative model of care, the physician is a teacher or friend who uses dialog to engage with the patient on the best treatment option [23]. He/she helps the patient with moral self-development and, in this way, empowered to consider all health-related values and their worthiness as related to implications for treatment.

Each of these models involve aspects of patient autonomy and purport a different degree of shared decision-making. Certain models may be more appropriate in specific clinical situations than others, or for those with particular patient characteristics. However, [21] argue that the deliberative model is the ideal physician-patient relationship when implemented effectively. This deliberative model, in fact, espouses the concepts of autonomy, empowerment, and SDM that constitute the Gothenburg model of person-centred care, which focuses on co-creation of care through partnership [17].

Although the active ingredients of the Gothenburg Person Centered Care (PCC) model have not been studied separately, a large body of scientific evidence across various settings in Sweden demonstrate the relationship between PCC and care outcomes [6]. Deliberative SDM specifically, then, maybe contributing to these positive results.

As alluded to previously, SDM has historically been a heterogeneous concept. Therefore, it is not surprising that conceptual models linking SDM to health outcomes are lacking in the literature. Various models do exist to explain the potential relationship between SDM, as a form of physician-patient communication, and health outcomes. The model by Street and colleagues [56] is of particular relevance. Using a broad definition of health outcomes that includes both physical health and emotional well-being, both direct and indirect pathways linking physician-patient communication to health outcomes is hypothesized. Applying this model to deliberative SDM in the context of PCC, SDM might enhance care outcomes via one or two indirect paths:

(1)   via proximal outcomes of the care interaction on the patient, such as increased satisfaction with the encounter, understanding of condition and options, trust in the care provider, feeling recognized/validated/heard/known, feeling involved in care decisions, motivation to take responsibility for own care, and/or
(2)   via proximal outcomes leading to intermediate outcomes affecting the patient, potentially including improved access to needed care, quality medical decisions affecting care, commitment to treatment, trust in the system in which care is received, social support, and selfcare skills.

Shay et al. [55] adapted this model to incorporate elements from the Transformation Model of Communication and Health Outcome by Kreps and colleagues [32]. This model describes how health communication, including SDM, might impact various aspects of the patient. Physician-patient communication might influence cognitive-affective components (e.g., trust, satisfaction), behavioral components (e.g., adherence, adoption of health behaviors), and/or physiological components (broadened to include self-rated health, quality of life, and clinical indicators such as blood pressure). This model was found to be an especially useful heuristic to synthesize the existing literature on SDM and health outcomes.

The number of scientific publications examining SDM's use, effectiveness, and relationship with health/care outcomes has increased exponentially in the past decade. However, the first published systematic review on patient outcomes and SDM occurred in 2015. Shay et al. [55] review articles with shared decision making in the title published in 2012 or prior. To be included in the review, all articles had to collected data on:

(1)   at least one perspective of SDM: from the patient self-report, clinician self-report of using SDM with patients, or observer-ratings of the use of SDM (e.g., via structured qualitative coding of audio-recorded encounters
(2)   at least one patient outcome: affective-cognitive, behavioral, or health.

Forty-one articles from 39 unique studies across various clinical contexts were identified for synthesis. The large majority of studies measured SDM via patient self-report ($n = 33$; 85%), while 15% ($n = 6$) used observer rating, and only 2 (8%) used clinician self-report.

Regarding patient outcomes, 97 different assessments were categorized with just over half as affective-cognitive (51%; $n = 50$ with half of these examining patient satisfaction), 28% were behavioral ($n = 27$), and slightly more than a fifth were health (21%; $n = 20$) outcomes.

Of the 97 relationships between SDM and an individual patient outcome, less than half (n = 42; 43%) were statistically significant. They created a 3 × 3 table and examined the number of significant relationship by SDM measurement perspective and type of outcome. Significant associations were most common (52%) when the patient reported SDM occurred, while only 21% of associations were significant when SDM was observer-rated and 0% when clinician-reported. Significant associations were also most common (54%) when affective-cognitive outcomes were examined, while only 37% of associations were significant when outcomes were behavioral and 25% when they were healthy. It is important to note that negative associations between SDM and patient outcomes were also noted in three articles. In summary, when the patient feels involved in SDM, they are also more likely to report more trust in their provider and satisfaction with care.

The former systematic review suggests that much more nuanced research into the potential relationships between SDM and patient/care outcomes is required to enhance understanding into potential mechanisms of action. In particular, additional emphasis on non-self-reported outcomes is warranted. In this respect, it is important to note that over 20 studies in Sweden based on 15 controlled clinical trials in 11 different disease/clinical areas with 2,610 people have been conducted examining the impact of the Gothenburg model of PCC on patient outcomes [6]. These findings suggest that PCC, which includes SDM, can impact patient outcomes in ways that go beyond individual perception. Such outcomes include lower gestational weight gain [25], shorter lengths of hospital stay [18, 46, 47] and cost savings [26]. While it is difficult to thease apart the active components of the Gothenburg PCC intervention, deliberative SDM can be considered to be closely associated with core PCC routines of creating, working, and maintaining a physician-patient/person partnership.

SDM concept could be consider from 3 perspectives: on the individual (micro) level, on the organisational (meso) level, as well on the policy-making (macro) level.

Individual (Micro) Level

The patient's benefit of SDM is that patients develop preferences based on their comprehension of accessible information. Patient/person has more realistic expectations, less "decisional conflict" as currently measured, and greater satisfaction. Shared decision-making can improve adherence and increase trust [10].

Individual-level effects on healthcare professional (HCP) should also be considered; the experience of supporting patients in arriving at informed decisions may be intrinsically rewarding. HCP may also find the effort involved emotionally and cognitively taxing, adding to their workload burden. These consequences for clinicians should be evaluated and understood as they should be expected to influence the uptake of shared decision-making [20].

Organizational (Meso) Level

The impact of shared decision-making as a communicative process has the enormous benefit on patients. But share decision making considers communication among the HCPs, as well as pharmacists, nurses. It can facilitate interprofessional barriers. Experts in organizational psychology consider five types of potential outcomes: (a) tangible outcomes products of teaming, e.g., costs or rates; (b) attitudes or emergent states, e.g., trust, psychological safety; (c) cogni-tive states, e.g., shared mental models; (d) team behaviours, e.g., turnover or absenteeism; and (e) norms, e.g., expected behaviours [20].

Policy-Making (Macro) Level

Shared decision-making has been welcomed by policy-makers world wide—it resonates and supports the ethical imperative of respect for patient autonomy and engagement [20].

One of the most cited SDM models that reflects all levels of the healthcare system is the **Interprofessional shared decision-making model** (IP-SDM) proposed by Légaré et al. [33].

In order to talk about an interprofessional approach to SDM, the IP-SDM model must include at least two health professionals of different professional orientations who either simultaneously or in various phases cooperate with each other in the SDM process with the patient. Communication on the individual (micro) level considers 6 steps.

Step 1: First step presents the patient with the health problem and requires a decision. A counterweight is a situation in which there is a decision point with more than one option and for which it is necessary to "weigh" each option well—the advantages and the disadvantages for each.

Step 2: This step involves exchanging information regarding options relevant to the patient's health condition. Healthcare professionals and patients share information about the potential benefits and potential side effects of each of the options, using educational materials, patient decision aids, and other evidence-based sources. Patient decision aids are tools that help people involved in decision-making processes, providing all the details, ie. information on all the benefits and dangers that this decision brings with it. Likewise, the participants in the IP-SDM process, discuss about the options available.

Step 3: Requires valuation by SDM process participants. Although patient values are the most important for making the right decisions, this model recognizes all the values that all SDM participants add to the system. Also, future research should consider the impact of numerous value sets on the IP-SDM process.

Step 4: Emphasizes the need to consider the feasibility of each of the options during the SDM process. Given the different business conditions in different countries, and thus in health care systems, SDM options will also differ. For this reason, local expertise is definitely not trivial for the functioning of this conceptual IP-SDM model.

Step 5: In this step, the essential decision is made. With the help of various experts, the patient decides on the preferred option. Healthcare professionals may also have their preferred option to share with the patient in the form of their recommendation. Ideally, both the patient and healthcare professionals would agree on the final option. After that, health professionals would organize a whole set of measures and procedures

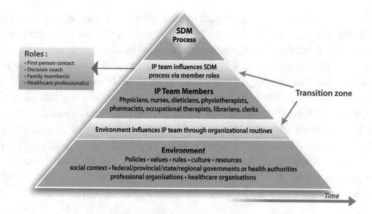

Fig. 5.1 IP-SDM model—healthcare system (meso and macro) levels representing the global influences in which the individual level is embedded (Légaré et al. 2010)

by which the patient can access the desired option. In case of disagreement, the final decision may be postponed.

Step 6: Involves patient support, so that the chosen option has the best impact on positive health outcomes, while achieving higher values (levels) of health service.

IP-SDM influence on the meso and macro levels is shown in Fig. 5.1.

At the top of the pyramid in the Fig. 5.1 represents the individual level of the process, which is explained earlier, trough 6 steps. Other parts of the pyramid depict either elements or persons from the health care system that may affect SDM. The dark shaded part in the middle, represents health professionals who can be involved in the SDM process (meso level). The dark shaded part at the bottom symbolizes the elements of the global environment—resources, government, cultural values, professional organizations and rules (macro level). Finally, the two transition zones represent the way in which elements of the health care system and individuals affect SDM. According to the top, the health care system acts on the processes of SDM through the rules implemented by its participants. For this reason, the team must develop mutual communication that is original, constructive and open, in order to foster mutual respect between SDM members, as well as between the team and the patient. At the lower level of the pyramid, the influence of the global environment on the organization of the team and on its functioning is explained.

The model also highlights the impossibility of functioning of interprofessional SDM outside the influence of factors at all levels of health care. Thus, within health teams (meso level), the interprofessional approach to SDM is influenced by the professional role of each individual member, and each professional role of the member is nurtured or limited by organizational routines or innovations within teams. Teams are also part of a larger organizational and social unit, which unite in a global environment (macro level). Despite the existence of good cooperation within IP-SDM teams, they necessarily require the support of the government, its policies and the leaders of health system organizations, as well as the sharing of a common goal. If there is no harmonization of common goals at all levels of the graphically represented pyramid, the implementation and successful functioning of the IP-SDM model is immediately disabled. Thus, the common goal is a *condition sine qua non* of the successful functioning of the entire system of IP-SDM models. Without a common understanding between health professionals, regulators, decision-makers, public opinion, and other stakeholders, there are inevitable difficulties in both effective communication and an understanding of the common goal across all levels of competence. Leger et al. revised this original IP-SDM model and validated them in primary care. Revision considered terminology change: instead "patient"—more convenient term "person". The revised model merges the micro, meso and macro levels in an integrated version that can help inform an IP approach to SDM in primary care [34].

The latest systematic review about SDM in health care included 50 articles and each describing a unique SDM model [27]. Twelve models were generic, the others were specific to a healthcare setting. All models consistently share some/same components: *Make the decision, Patient preferences, Tailor information, Deliberate, Create choice awareness,* and *Learn about the patient*. The overall conclusion was that a unified view on what SDM is still lacking.

Critical ingredients of SDM, and their overlap with PCC, could be further explored. Such work could help to advance the field of SDM by contributing new aspects of SDM measurement and potentially enhance understanding of the link between SDM and objectively measured patient outcomes.

5.3 Multicriteria Decision Making Techniques and Shared Decision Making

As stated in the previous part, shared decision making (SDM) describes a collaborative process in which healthcare providers and patients/families make treatment decisions using the best available evidence, while taking into account the patient's values and preferences.

If we take into account person-centred care as one of the most important frameworks to improve patient outcomes, in which the same parties as in SDM are involved, we come to the urge for techniques and methods that can support such a complex decision-making process.

This decision-making process is complex on its own because many parties (patients, healthcare professionals, reimbursement funds, health authorities, policymakers, etc.) with different perspectives and approaches are involved, and it is getting even more complex with the evolution of evidence-based medicine with new "big data" and "real-world evidence (RWE)" approaches in order to gain the best scientific evidence currently available.

If we want to have implementation of SDM in the best way, we need to have access to current evidence comparing expected outcomes of decision alternatives, assessment of decision-related values and preferences, and integration of this information to identify the most suitable course of action. Since decisions in healthcare settings are not unique and significantly different from other areas, methods that are well-known and are in use for many decades can be utilized.

Multi-criteria decision analysis (MCDA) methods fulfil all of the required elements of SDM, and this suggests that MCDA methods could be used effectively to facilitate SDM in practice [42].

Multiple criteria-decision making (MCDM) refers to making decisions in the presence of multiple, usually conflicting criteria, allowing with its methods involvement of different (also often conflicting) stakeholders' perspectives, preferences and values, which are then mutually compared, analysed and unique decision is made which represent the best compromise solution for everyone involved in decision making process.

Benjamin Franklin could be one of the first advocate of MCDA as he was using a paper-based system when making important decisions. He would write down the arguments for and against one decision on different sides of the paper, after that he would strike out those arguments on each side of the paper that had relatively equal importance, and when all arguments were eliminated on one side, he would look on the other side, if there were no arguments left, he would make decision [15].

Multi-criteria analysis (MCDA) is a general term that includes a number of analytical techniques used in the decision-making process in the context of multiple, and often conflicting criteria. These techniques serve to support decision-makers on how to agree on which evaluation criteria are relevant, how important these criteria are, and how this information can be used as an alternative procedure (option). MCDA encompasses a broad set of methodological approaches, derived from operational research but with a rich intellectual foundation in other disciplines [42].

Process of MCDA could be simply described in following way: first, identify interventions to evaluate, then identify criteria against which to evaluate the interventions, then measure the interventions against the criteria and at the end, combine the criteria scores to produce a ranking of each intervention [15].

Often, in literature, many terms can be found that are used interchangeably without universal definitions of them, we suggest when screening literature for MCDM, specially in healthcare, you should include all of the following terms: "multiple objective", "multiple criteria", "multiple attribute", "multi-objective", "multi-criteria", "multi-criteria decision analysis (MCDA)", "multi-attribute decision making (MADM)", "multi-objective decision making (MODM)", "multi-attribute utility theory (MAUT)". We think that distinguish should be made, where differences in practise exists, and that is of high interest that universal definitions of these terms in healthcare settings and consensus about them are made. In this chapter, we will use MCDA, MCDM and MODM as synonyms.

The use of MCDA in health systems is on the rise, a number of published papers in previous years are growing rapidly, and it should be seen as a natural continuation of evidence-based medicine assessment and RWE implementation. The MCDA also provides a set of techniques for determining which performance elements (criteria) need to be measured, what is their importance, how stakeholder preferences can be expressed, and how performance data and preferences can be combined to assess alternatives [42].

The challenge for users of the MCDA method is that a large number of different MCDA techniques are available, and there are few guidelines available on how to decide which technique to choose from all available. For model building most frequently used methods are: Value measurement models (Multi-attribute Value Theory, Multi-attribute Utility Theory, Analytical Hierarchy Process), Goal Programming, Reference models, Outranking model (ELECTRE method, PROMETHEE method, GAIA method). The four most commonly used MCDA methods are [42]: Direct rating, Keeney-Raiffa MCDA, Analytical hierarchical process and Discrete choice experiment.

All of these methods have its advantages and disadvantages, and carefully method should be chosen from case to case. Many of these techniques are already being used in healthcare decision making process, but on the other hand many of them yet have not been, and further research of implementation of these methods in practise using real world data in SDM should be done.

Marsh et al. illustrated how two MCDA methods—the conjoint analysis and analytic hierarchy process (AHP)—have been used to foster shared decision-making in clinical settings [42].

The importance of MCDA in a healthcare setting is evidenced by the EVIDEM framework which promotes transparent and efficient healthcare decision-making and provides a collaborative framework [24] and also by the fact that ISPOR established a working group for new good practices for MCDA ("Emerging Good Practices Task Force for MCDA") in 2014, and since then two reports have been published which define MCDA in healthcare settings, provide guidelines for conducting it, procedural values of MCDA and basic steps to be followed [43, 57].

Examples of countries where some of the MCDA techniques have been applied to support health decision making are: Canada, Germany, Lombardy in Italy, Hungary, South Africa, Thailand, New Zealand and many others [14]. Recently published article suggests possible use of MCDA in benefit-risk framework, for European Medical Agency (EMA) and Food and Drug Administration (FDA), and mentions EMA's experience in the IMI-PROTECT projects [2].

Many examples of MCDA use in healthcare settings which involves SDM can be found, only some of them (as a representation of its possible diversity applications) which are published in 2020, are given in the Table 5.1.

There is broad spectre of MCDA use in healthcare, especially in SDM process, on different levels: from individual level (micro level)—the best therapy choice for one patient; meso level—prioritization for hospital admission of patients in one city; to macro level—evaluation of new technology (not) to be reimbursed, guidelines development, prioritization of the dossier evaluation at regulatory authorities.

The best way to explore advantages and disadvantages of each technique and to make them more visible and acceptable to decision-makers on macro-level is to fund and promote: development of global guidelines and consensus on defining

Table 5.1 Selected articles of MCDA use in healthcare setting in 2020

Article name	Author
Multi-Criteria Decision Analysis to prioritize hospital admission of patients affected by COVID-19 in low-resource settings with hospital-bed shortage	De Nardo et al. [12]
Multi–Criteria–Decision–Analysis (MCDA) for the Horizon Scanning of Health Innovations an Application to COVID 19 Emergency	Ruggeri et al. [49]
Early Health Technology Assessment during Nonalcoholic Steatohepatitis Drug Development: A Two-Round, Cross-Country, Multicriteria Decision Analysis	Angelis et al. [3]
Comprehensive value assessment of drugs using a multi-criteria decision analysis: An example of targeted therapies for metastatic colorectal cancer treatment	Hsu et al. [28]
A methodology based on multiple criteria decision analysis for combining antibiotics in empirical therapy	Campos et al. [8]
Development of a Multicriteria Decision Analysis Framework for Evaluating and Positioning Oncologic Treatments in Clinical Practice	Camps et al. [9]
Assessment and prioritization of the WHO "best buys" and other recommended interventions for the prevention and control of non-communicable diseases in Iran	Bakhtiari et al. [5]
Benefit and risk of Tripterygium Glycosides Tablets in treatment of rheumatoid arthritis based on multi- criteria decision-making analysis	Jiang et al. [29]
Multiple criteria decision analysis approach to consider therapeutic innovations in the emergency department: The methoxyflurane organizational impact in acute trauma pain	Lvovschi et al. [39]
Feasibility of Measuring Preferences for Chemotherapy Among Early-Stage Breast Cancer Survivors Using a Direct Rank Ordering Multicriteria Decision Analysis Versus a Time Trade-Off	Panattoni et al. [48]
Assessing the Preferences for Criteria in Multi-Criteria Decision Analysis in Treatments for Rare Disease	Schey et al. [54]
'It takes two to tango': Bridging the gap between country need and vaccine product innovation	Archer et al. [4]

terms, wording and how to conduct MCDA in healthcare settings; implementation of different MCDA techniques as much as possible; mutual comparison of different MCDA methods on the same topic. The most important is to involve in all these researches not only all interested parties rather all possibly involved and impacted parties in decision making process (eg. patient representatives, regulators, policymakers).

The first book related to this topic and its importance is "Multi-Criteria Decision Analysis to Support Healthcare Decisions" published in 2017 [42], and we strongly recommend it to anyone who wants to step into the amazing world of MCDA.

5.4   Ethical and Practical Consideration About Shared Decision-Making in Person-Centered Care

The focal point of person-centered care (PCC) is treating patients in a broad context, not only through the prism of their illnesses and biomedical tests, but most of all perceiving them as persons with all their capabilities and limitations [17, 38]. PCC means moving from a model where patients are passive targets for medical intervention to a model where more partner-like arrangements are made involving the patients as active partners in their care. Co-creation of care and partnership between patients, their families, and healthcare professionals is an essential element of PCC. The GPCC model of PCC entails three pillars consisting of: (1) initiating a partnership by inviting the patient to narrate about their daily life in relation to their condition, the sick person's description of their illness, symptoms and their impact on their life; (2) the process of shared decision-making, based on the unique narrative of the patient and the generic knowledge of the professional; (3) the process of safeguarding the partnership by documenting the sick person's narrative and a jointly agreed care plan that is regularly reviewed and updated [6]. The above brief description indicates that PCC is a very complex approach in many points overlapping with shared decision-making (SDM).

The Patient Narrative in SDM

Central notions in PCC and SDM is patient narrative. The notion 'patient narrative' suggests a rich base of information about a patient, in addition to biomedical tests and physical examination. The information delivered by patients through the 'narration' promises complexity and multidimensionality of the information, including not only patient's values and preferences but also experiences and wants that have to be included in clinical decision making. PCC and SDM call for a patient narrative to be 'holistic', which means to embrace all aspects of the patient's situation, from biomedical, over psychological to social and existential aspects or problems. Some of them might be sensitive details about personal feelings, social relationships, and embarrassing aspects. The 'holistic' approach, however, may bring many ethical and practical issues that should be considered in order to provide more benefits than harm through using SDM in PCC [45].

Early theorists of SDM, believed that safeguarding a deliberative dialogue between caregivers and patients would by itself protect against paternalism and promote patient autonomy in a way undoubtedly required from the ethical perspective [21]. The practice, however, is far more complex and challenging [11, 52]. It has to be noted that SDM in healthcare is significantly different than in other areas, for example, in commercial business. In a business partnership, partners may be coequal in terms of knowledge and power, and bear the same consequences of the decisions made. In the medical encounter, professionals know medical facts that patients do not, patients, on the other hand, have an experience of illness that is unique and probably never was and will be an experience of professionals, and patients, not professionals, will bear the consequences of the shared clinical decisions [58]. These create inescapable asymmetries and discrepancy of interests.

PCC and SDM assume professionals' competencies to comprehend such 'holistic' and multi-dimensional problems [41], as well as the capability to identify and record the most relevant details in order other professionals, could use it. The documentation is especially important when the PCC approach is implemented in the whole chain of treatment in the medical organization, and patients suffer from several different illnesses what is common in an older population. In such circumstances documented patient narrative should follow the patient route through different professionals and providers, that the patient does not need to repeat the narrative again and again.

However, it has to be acknowledged that the description of the patient narrative recorded in clinical documentation may be biased. Tonelli and Sullivan [58] suggest that clinician values and subjective judgments may enter into the clinical relationship before the diagnosis and negotiations determining what symptoms and features are the most important to investigate and highlight in the documentation. Any attempt to oblige clinicians to provide unbiased information about patients' narrative is doomed to fail. Depriving professionals of including their values in their clinical work is inconsistent with recognizing also professionals within the clinical encounter in the SDM paradigm [44]. In the PCC approach, SDM is an ongoing process within which the patient narrative can be continuously added, revised, and interpreted. This means that time is needed for the narrators to deliver their stories and the listener to understand and analyse them. Therefore, managers of medical providers have to ensure that professionals have access to adequate resources [36] to manage patients' narrations appropriately [41].

Boundaries Within Which PCC and SDM Brings Benefits to Patients

SDM concept had some barriers for implementation.

Patients generally do not expect SDM in the health care system, nor do they have any objections to its absence. They often feel powerless and submissive when meeting with health care providers. Also, the lack and limited time is the most frequently cited reason by doctors for the impossibility of successful implementation of SDM concept in everyday practice. Although SDM does require more time spent in consultation with patients, studies have not identified any link between improving the level of communication with extending the time spent specifically on communication [33].

Documentation and distribution of the patient narrative may benefit as well as harm patients. Information from patient narrative could be used in decision making in ways that promote and protect patient autonomy, as well as in a paternalistic manner [52]. Wider knowledge about patients may empower professionals to better assist patients in clarifying their own ideas and decisions about their care in the context of their life. However, this knowledge may also provide wider opportunities for manipulating, confusing and interfering with patients' decision making in ways that reduce their autonomy. Interfering with patients' decisions may not be the result of a purposeful paternalist strategy of professionals, but maybe the result of a lack of their skills and knowledge about how to manage the patient narrative and dialogue needed for shared decision making [19]. This may increase the risk of professionals' mismanagement the SDM so that the patient is harmed rather than benefited. For example, when the knowledge about medical aspects conveyed by professionals to patients would be too technical and difficult to comprehend, patients could become blocked in their decision making.

Wide implementation of PCC and consequently SDM in medical providers may make patients feel pressed to talk more about their personal life, intimate feelings, or difficulties in social relationships than they would otherwise have liked, giving room to retrospective regret and unease and affect trust in professionals [35]. A patient's broad narrative about their personal problems may also emotionally burden professionals, especially when they would be untrained or feel a lack of organizational support. Thus, expanding the practice of SDM in an organization calls for the implementation of monitoring and evaluation to assure that the benefits are not accompanied by harmful effects [45].

The problem is that in practice, during the implementation of PCC and SDM, professionals are rather not trained on how to handle and prioritize knowledge about patients' lives. This issue is also not mentioned in the WE-CARE Roadmap, a framework supporting the implementation of PCC [37]. Similarly, not all patients are prepared to adequately match treatment options to personal preferences and values, taking into account rather long-lasting effects than immediate well-being. Thus, effective implementation of PCC, require to follow the narrow space between patients' sufficient narrative and their capacity of autonomy and between professionals' skills to handle the narrative and their ability to present adequate treatment options in a way understandable to patients.

Professionals may select and document information from the narrative in a way that may promote either paternalist or autonomy strategy. This may not only depend on professionals' attitude but also on professionals' subjective judgment about patients' capability to make autonomous decisions, which is apparently a matter of a continuous range [51]. But to support the subjective judgment of professionals about patients' decision-making competency, theoretical or ethical recommendations have to be established. Although from the above discussion appears that SDM would be most suitable for decision-competent patients, the practice of PCC seems to regard patients with significant cognitive decline, such as patients with dementia (e.g. [13, 16]. In PCC this decision incompetency is resolved by the inclusion of other people, such as adult children of demented elderly people. However, the inclusion of other people in SDM raise questions about whether adult children are more competent to make actually paternalistic decision then professionals? And whether the interests of the family that take care of the cognitively impaired person are in concordance with the interests of the person? There are also other questions related to the benefits of PCC and a person's autonomy. For example, whether relocating decisional authority to decision incompetent patients may foster them to become more competent or improve their compliance or adherence [53]?

There is also another issue, that not all patients wish to talk much about themselves and engage much in decision making [1], thus their inclusion in PCC:

'seems to imply a sort of meta-paternalism: the idea that patients need to engage in PCC/SDM procedures to access further care seems to mean forcing, pressing or manipulating some people to engage in (allegedly) autonomy and/or health-promoting procedures against their wishes (a sort of paternalism that would remain even if the goal of PCC/SDM was assumed to be entirely focused on autonomy promotion). This highlights further the lack of basic ethical clarity regarding what are supposed to be the point and limits of PCC/SDM in autonomy and paternalism terms. It also makes salient the existence of different autonomy ideals that PCC/SDM may be related to.' [45].

One more narrow passage that PCC practitioners must navigate is the shift from compliance to adherence in relation to the paradigm of evidence-based medicine (EBM). While compliance is the following of doctor's instruction who are obliged to prescribe the best possible therapy according to EBM, adherence is a patients' proactive behavior capturing the idea of professionals permitting patients to take part in clinical decision making and to correct professional clinical judgment following patients' preferences, values and wants. Thus, adherence as an ingredient of PCC and SDM strengthens patients' influence of the treatment plan, which at the starting point should be the best option according to EBM. This raises the ethical issue to what extent professionals may accept patients' autonomy in decision-making about treatment plans knowing that patients' demands may compromise clinical outcomes. What about professional responsibility and the Hippocratic Oath? How far the professionals can go in terms of persuasion, incentives, and pressures to influence patients' decisions and still maintain an adequate level of patients' autonomy. The level of autonomy required for decision making to be truly shared. This means that *'there must be agreement between participants regarding all aspects of clinical choice. Simply engaging in the process of consultation does not constitute sharing decisions.'* [58].

Munthe et al. [45] draw attention to the problem, that in PCC and SDM, the key component is continuity, and therefore adherence in relation to mutually agreed treatment plan may not be secured by one agreement. This means that the treatment plan may be actualized during the new round of SDM, leading to a new agreement where issues making the patient non-adherent are taken into consideration to secure future adherence. However, maybe further from the optimal, based on EBM, treatment plan. This raises the question of to what extent patients not having deep clinical knowledge may take responsibility for their treatment? Whether at the end patients would not regret their own decisions and blame professionals for not convincing them to the EBM treatment plan. This is a very important issue taking into account above mentioned limitations regarding the abilities to comprehend patients' narratives by professionals and professional recommendations by patients.

Tonelli and Sullivan [58] claim that in SDM professionals and a patient have to choose the same thing for the patient. When there is a discrepancy, it could mean that the professionals do not understand the patient or that the patient does not understand what the professionals are trying to convey. Within a medical relationship, such discrepancy should be an invitation to a further examination of the knowledge and dialog since in SDM, it is not sufficient to accept the patient's choice. Professionals have to really understand and accept the patient's rationale for the choice. According to [58], when there is intractable disagreement, it should lead to the end of the medical relationship between the patient and professionals.

5.5 Conclusion

This chapter examined shared-decision (SDM) models as an approach to person-centered care (PCC). SDM concept could be recognised in three levels (micro, meso, macro) of implementation that we introduced.

Multi-criteria decision analysis (MCDA) methods could be used effectively to facilitate SDM in practice, and different techniques in the recent publications were presented. The chapter concluded with an overview of the benefits and barriers in shared-decision implementation in the healthcare system. SDM can be closely associated with core PCC routines of creating, working, and maintaining a physician-patient/person partnership.

Acknowledgements This publication is based upon work from COST Action "**European Network for cost containment and improved quality of health care-CostCares**" (CA15222), supported by COST (European Cooperation in Science and Technology)

COST (European Cooperation in Science and Technology) is a funding agency for research and innovation networks. Our Actions help connect research initiatives across Europe and enable scientists to grow their ideas by sharing them with their peers. This boosts their research, career and innovation.

https://www.cost.eu

Funded by the Horizon 2020 Framework Programme of the European Union

References

1. Ågård, A., Löfmark, R., Edvardsson, N., Ekman, I.: Views of patients with heart failure about their role in the decision to start implantable cardioverter–defibrillator treatment: prescription rather than participation. J. Med. Ethics **33**(9), 514–518 (2007). https://doi.org/10.1136/jme.2006.017723
2. Angelis, A., Phillips, L.: Advancing structured decision-making in drug regulation at the FDA and EMA. Br. J. Clin. Pharmacol. 1–11 (2020). https://doi.org/10.1111/bcp.14425
3. Angelis, A., Thursz, M., Ratziu, V., O'Brien, A., Serfaty, L., Canbay, A., Schiefke, I., Costa, J., Lecomte, P., Kanavos, P.: Early health technology assessment during nonalcoholic steatohepatitis drug development: a two-round, cross-country, multicriteria decision analysis. Med. Decis. Making **40**(6), 830–845 (2020). https://doi.org/10.1177/0272989X20940672
4. Archer, R.A., Kapoor, R., Isaranuwatchai, W., Teerawattananon, Y., Giersing, B., Botwright, S., Luttjeboer, J., Hutubessy, R.: "It takes two to tango": Bridging the gap between country need and vaccine product innovation. PLoS ONE **15**(6), e0233950 (2020). https://doi.org/10.1371/journal.pone.0233950

5. Bakhtiari, A., Takian, A., Majdzadeh, R., et al.: Assessment and prioritization of the WHO "best buys" and other recommended interventions for the prevention and control of non-communicable diseases in Iran. BMC Public Health **20**, 333 (2020). https://doi.org/10.1186/s12889-020-8446-x

6. Britten, N., Ekman, I., Naldemirci, Ö., Javinger, M., Hedman, H., Wolf, A.: Learning from Gothenburg model of person centred healthcare. BMJ **370**, m2738 (2020). https://doi.org/10.1136/bmj.m2738

7. Burke, G.: Ethics and medical decision-making. Prim. Care **7**(4), 615–624 (1980)

8. Campos, M., Jimenez, F., Sanchez, G., Juarez, J.M., Morales, A., Canovas-Segura, B., Palacios, F.: A methodology based on multiple criteria decision analysis for combining antibiotics in empirical therapy. Artif. Intell. Med. **102**, 101751 (2020). https://doi.org/10.1016/j.artmed.2019.101751

9. Camps, C., Badia, X., García-Campelo, R., García-Foncillas, J., López, R., Massuti, B., Provencio, M., Salazar, R., Virizuela, J., Guillem, V.: Development of a multicriteria decision analysis framework for evaluating and positioning oncologic treatments in clinical practice. JCO Oncol. Pract. **16**(3), e298–e305 (2020). https://doi.org/10.1200/JOP.19.00487

10. Carman, K.L., Dardess, P., Maurer, M., Sofaer, S., Adams, K., Bechtel, C., Sweeney, J.: Patient and family engagement: a framework for understanding the elements and developing interventions and policies. Health Aff. **32**(2), 223–231 (2013). https://doi.org/10.1377/hlthaff.2012.1133

11. Clarke, G., Hall, R.T., Rosencrance, G.: Physician-patient relations: no more models. Am. J. Bioeth. **4**(2), W16-19 (2004). https://doi.org/10.1162/152651604323097934

12. De Nardo, P., Gentilotti, E., Mazzaferri, F., Cremonini, E., Hansen, P., Goossens, H., Tacconelli, E., Members of the COVID-19 MCDA Group.: Multi-criteria decision analysis to prioritize hospital admission of patients affected by covid-19 in low-resource settings with hospital-bed shortage. Int. J. Infect. Dis. **98**:494–500 (2020). https://doi.org/10.1016/j.ijid.2020.06.082

13. Dichter, M. N., Reuther, S., Trutschel, D., Köpke, S., Halek, M. (2019). Organizational interventions for promoting person-centred care for people with dementia. Cochrane Database Syst. Rev. 2019(7):CD013375.:https://doi.org/10.1002/14651858.CD013375

14. Drake, J.I., de Hart, J.C.T., Monleon, C., Toro, W., Valentim, J.: Utilization of multiple-criteria decision analysis (MCDA) to support healthcare decision-making FIFARMA, 2016. J. Mark Access Health Policy **5**(1), 1360545 (2017). https://doi.org/10.1080/20016689.2017.1360545

15. Edwards and McIntosh: Applied Health Economics for Public Health Practise and Research, pp. 301–311. Oxford University Press, United Kingdom (2019)

16. Edvardsson, D., Sandman, P.O., Borell, L.: Implementing national guidelines for person-centered care of people with dementia in residential aged care: effects on perceived person-centeredness, staff strain, and stress of conscience. Int. Psychogeriatr. **26**(7), 1171–1179 (2014). https://doi.org/10.1017/S1041610214000258

17. Ekman, I., Swedberg, K., Taft, C., Lindseth, A., Norberg, A., Brink, E., Carlsson, J., Dahlin-Ivanoff, S., Johansson, I.L., Kjellgren, K., Lidén, E., Öhlén, J., Olsson, L.E., Rosén, H., Rydmark, M., Sunnerhagen, K.S.: Person-centered care—Ready for prime time. Eur. J. Cardiovasc. Nurs. **10**(4), 248–251 (2011). https://doi.org/10.1016/j.ejcnurse.2011.06.008

18. Ekman, I., Wolf, A., Olsson, L.E., Taft, C., Dudas, K., Schaufelberger, M., Swedberg, K.: Effects of person-centred care in patients with chronic heart failure: the PCC-HF study. Eur. Heart J. **33**(9), 1112–1119 (2012). https://doi.org/10.1093/eurheartj/ehr306

19. Elwyn, G., Edwards, A., Gwyn, R., Grol, R.: Towards a feasible model for shared decision making: focus group study with general practice registrars. BMJ **319**(7212), 753–756 (1999). https://doi.org/10.1136/bmj.319.7212.753

20. Elwyn, G., Frosch, D., Kobrin, S.: Implementing shared decision-making: consider all the consequences. Implement Sci. **11**, 114 (2016). https://doi.org/10.1186/s13012-016-0480-9

21. Emanuel, E.J., Emanuel, L.L.: Four models of the physician-patient relationship. JAMA **267**(16), 2221–2226 (1992). https://doi.org/10.1001/jama.1992.03480160079038

22. Farrell, C., Towle, A., Godolphin, W.: Where's Patient's Voice in Health Professional Education? University of British Columbia, Vancouver, BC (2006)

23. Fried, C.: The lawyer as friend: the moral foundations of the lawyer client relationship. Yale L J. **85**, 1060–1089 (1976)
24. Goetghebeur, M.M., Wagner, M., Khoury, H., Levitt, R.J., Erickson, L.J., Rindress, D.: Evidence and value: Impact on DEcisionMaking-the EVIDEM framework and potential applications. BMC Health Serv. Res. **8**, 270 (2008). https://doi.org/10.1186/1472-6963-8-270
25. Haby, K., Glantz, A., Hanas, R., Premberg, Å.: Mighty mums—An antenatal health care intervention can reduce gestational weight gain in women with obesity. Midwifery **31**(7), 685–692 (2015). https://doi.org/10.1016/j.midw.2015.03.014
26. Hansson, E., Ekman, I., Swedberg, K., Wolf, A., Dudas, K., Ehlers, L., Olsson, L.E.: Person-centred care for patients with chronic heart failure—A cost-utility analysis. Eur. J. Cardiovasc. Nurs. **15**(4), 276–284 (2016). https://doi.org/10.1177/1474515114567035
27. Herlitz, A.: Comparativism and the grounds for person-centered care and shared decision making. J. Clin. Ethics **28**(4), 269–278 (2017)
28. Hsu, J.C., Lin, J.Y., Lin, P.C., Lee, Y.C.: Comprehensive value assessment of drugs using a multi-criteria decision analysis: an example of targeted therapies for metastatic colorectal cancer treatment. PLoS ONE **14**(12), e0225938 (2019). https://doi.org/10.1371/journal.pone.0225938
29. Jiang, H., Zhang, X.M., Zhang, B., Zhang, D., Lyu, J.T.: Benefit and risk of tripterygium glycosides tablets in treatment of rheumatoid arthritis based on multi-criteria decision-making analysis. Zhongguo Zhong Yao Za Zhi **45**(4), 798–808 (2020). https://doi.org/10.19540/j.cnki.cjcmm.20191115.502
30. Kaplan, R.M., Frosch, D.L.: Decision making in medicine and health care. Annu. Rev. Clin. Psychol. **1**, 525–556 (2005). https://doi.org/10.1146/annurev.clinpsy.1.102803.144118
31. Kerssens, J.J., Groennwegen, P.P., Sixma, H.J., Boerma, W.G., van der Eijk, I.: Comparison of patient evaluations of health care quality in relation to WHO measures of achievement in 12 European countries. Bull. World Health Organ. **82**(2), 106–114 (2004)
32. Kreps, G.L., O'Hair, D., Clowers, M.: The influences of human communication on health outcomes. Am. Behav. Sci. **38**, 248–256 (1994)
33. Légaré, F., Ratte, S., Gravel, K., Graham, I.D.: Barriers and facilitators to implementing shered decision-making in clinical practice: update of a systematic review of health professionals' perceptions. Patient Educ. Couns. **73**(3), 526–535 (2008). https://doi.org/10.1016/j.pec.2008.07.018
34. Légaré, F., Stacey, D., Pouliot, S., Gauvin, F.P., Descroches, S., Kryworuchko, J., Dunn, S., Elwyn, G., Frosch, D., Gagnon, M.P., Harrison, M.B., Pluye, P., Graham, I.D.: Interprofesionalism and shared decision-making in primary care a stepwise approach towards a new model. J. Interprof. Care **25**(1), 18–25 (2011). https://doi.org/10.3109/13561820.2010.490502
35. Lewandowski, R. A.: Trust in health care: susceptibility to change. In: Chodorek, M. (ed) Organizational relations as a key area of positive organizational potential. Wydawnictwo Naukowe Uniwersytetu Mikołaja Kopernika, Toruń, pp. 9–25 (2011)
36. Lewandowski, R.A.: Cost control of medical care in public hospitals—A comparative analysis. Int. J. Contemp. Manag. **13**(1), 125–136 (2014)
37. Lewandowski, R.: The WE-CARE roadmap: a framework for implementation of person-centred care and health promotion in medical organizations. J. Appl. Manag. Invest. **9**(3), 120–132 (2020)
38. Lloyd, H.M., Ekman, I., Rogers, H.L., Raposo, V., Melo, P., Marinkovic, V.D., Buttigieg, S.C., Srulovici, E., Lewandowski, R.A., Britten, N.: Supporting innovative person-centred care in financially constrained environments: the WE CARE exploratory health laboratory evaluation strategy. IJERPH **17**(9), 3050 (2020). https://doi.org/10.3390/ijerph17093050
39. Lvovschi, V.E., Maignan, M., Tazarourte, K., Diallo, M.L., Hadjadj-Baillot, C., Pons-Kerjean, N., et al.: Multiple criteria decision analysis approach to consider therapeutic innovations in the emergency department: the methoxyflurane organizational impact in acute trauma pain. PLoS ONE **15**(4), e0231571 (2020). https://doi.org/10.1371/journal.pone.0231571
40. MacIntyre, A.: After Virtue. University of Notre Dame Press, South Bend, Ind (1981)

41. Mead, N., Bower, P.: Patient-centredness: a conceptual framework and review of the empirical literature. Soc. Sci. Med. **51**(7), 1087–1110 (2000). https://doi.org/10.1016/s0277-9536(00)00098-8
42. Marsh, K., Goetghebeur, M., Thokala, P., Baltussen, R.: Multi-Criteria Decision Analysis to Support Healthcare Decisions. Springer International Publishing AG, Switzerland (2017)
43. Marsh, K., IJzerman, M., Thokala, P., Baltussen, R., Boysen, M., Kaló, Z., Lönngren, T., Mussen, F., Peacock, S., Watkins, J., Devlin, N., and ISPOR Task Force.: Multiple criteria decision analysis for health care decision making-emerging good practices: report 2 of the ISPOR MCDA emerging good practices task force. Value Health **19**(2), 125–137 (2016). https://doi.org/10.1016/j.jval.2015.12.016
44. Miles, A.: On a medicine of the whole person: away from scientistic reductionism and towards the embrace of the complex in clinical practice. J. Eval. Clin. Pract. **15**(6), 941–949 (2009). https://doi.org/10.1111/j.1365-2753.2009.01354.x. Dec
45. Munthe, C., Sandman, L., Cutas, D.: Person centred care and shared decision making: implications for ethics, public health and research. Health Care Anal. **20**(3), 231–249 (2012). https://doi.org/10.1007/s10728-011-0183-y
46. Olsson, L.E., Karlsson, J., Berg, U., Kärrholm, J., Hansson, E.: Person-centred care compared with standardized care for patients undergoing total hip arthroplasty—A quasi-experimental study. J. Orthop. Surg. Res. **9**, 95 (2014). https://doi.org/10.1186/s13018-014-0095-2
47. Olsson, L. E., Hansson, E., Ekman:. Evaluation of person-centred care after hip replacement - a controlled before and after study on the effects of fear of movement and self-efficacy compared to standard care. BMC Nurs. **15**(1), 53 (2016). https://doi.org/10.1186/s12912-016-0173-3
48. Panattoni, L., Phelps, C.E., Lieu, T.A., Alexeeff, S., O'Neill, S., Mandelblatt, J.S., Ramsey, S.D.: Feasibility of measuring preferences for chemotherapy among early-stage breast cancer survivors using a direct rank ordering multicriteria decision analysis versus a time trade-off. Patient **13**(5), 557–566 (2020). https://doi.org/10.1007/s40271-020-00423-w
49. Ruggeri, M., Cadeddu, C., Roazzi, P., Mandolini, D., Grigioni, M., Marchetti, M.: Multi-criteria-decision-analysis (MCDA) for the horizon scanning of health innovations an application to COVID 19 emergency. Int. J. Environ. Res. Public Health **17**(21), 7823 (2020). https://doi.org/10.3390/ijerph17217823
50. Sandel, M.J.: Liberalism and the Limits of Justice. Cambridge University Press, New York, NY (1982)
51. Sandman, L., Munthe, C.: Shared decision-making and patient autonomy. Theor Med Bioeth **30**(4), 289–310 (2009). https://doi.org/10.1007/s11017-009-9114-4
52. Sandman, L., Munthe, C.: Shared decision making, paternalism and patient choice. Health Care Anal. **18**(1), 60–84 (2010). https://doi.org/10.1007/s10728-008-0108-6
53. Segal, J.Z.: "Compliance" to "concordance": a critical view. J. Med. Humanit. **28**(2), 81–96 (2007). https://doi.org/10.1007/s10912-007-9030-4
54. Schey, C., Postma, M.J., Krabbe, P.F.M., Topachevskyi, O., Volovyk, A., Connolly, M.: Assessing the preferences for criteria in multi-criteria decision analysis in treatments for rare diseases. Front. Public Health **8**, 162 (2020). https://doi.org/10.3389/fpubh.2020.00162
55. Shay, L.A., Lafata, J.E.: Where is the evidence? A systematic review of shared decision making and patient outcomes. Med. Decis. Making **35**(1), 114–131 (2015). https://doi.org/10.1177/0272989X14551638
56. Street, R.L., Jr., Makoul, G., Arora, N.K., Epstein, R.M.: How does communication heal? Pathways linking clinician-patient communication to health outcomes. Patient Educ. Couns. **74**(3), 295–301 (2009). https://doi.org/10.1016/j.pec.2008.11.015
57. Thokala, P., Devlin, N., Marsh, K., Baltussen, R., Boysen, M., Kalo, Z., Longrenn, T., Mussen, F., Peacock, S., Watkins, J., Ijzerman, M.: Multiple criteria decision analysis for health care decision making—An introduction: Report 1 of the ISPOR MCDA emerging good practices task force. Value Health **19**(1), 1–13 (2016). https://doi.org/10.1016/j.jval.2015.12.003
58. Tonelli, M.R., Sullivan, M.D.: Person-centred shared decision making. J. Eval. Clin. Pract. **25**(6), 1057–1062 (2019). https://doi.org/10.1111/jep.13260
59. Veatch, R.M.: Models for ethical medicine in a revolutionary age. Hastings Cent Rep. **2**, 3–5 (1975)

Chapter 6
Advancement of Efficiency Evaluation for Healthcare

Fabien Canolle, Darijana Antonić, António Casa Nova, Anatoliy Goncharuk, Paulo Melo, Vítor Raposo, and Didier Vinot

Abstract The objective of this chapter is to provide conceptual understandings of evaluation methods for healthcare and concrete illustrations in order to take stock of the advancements and applications on the subject. The chapter is divided in four sections: the first one sets the stage at a European level by evaluating healthcare system performance; the second goes back to the fundamental principles of methods

F. Canolle (✉)
Univ. Grenoble Alpes, Grenoble INP*, CERAG, 38000 Grenoble Alpes Chaire Valeurs du soin, France
e-mail: fabien.canolle@univ-grenoble-alpes.fr

D. Vinot
Université de Lyon, Université Jean Moulin, Magellan research center, Chaire Valeurs du soin, France
e-mail: didier.vinot@univ-lyon3.fr

D. Antonić
Public Health Institute of the Republic of Srpska, Jovana Dučića 1, 78000 Banja Luka, Republic of Srpska, Bosnia and Herzegovina
e-mail: darijana.antonic@phi.rs.ba

A. C. Nova
Health School of Polytechnic Institute of Portalegre, Campus Politécnico 10, 7300-555 Portalegre, Portugal
e-mail: casanova@ipportalegre.pt

A. Goncharuk
Department of Management, International Humanitarian University, 33 Fontain Road, 65009 Odessa, Ukraine
e-mail: agg@ua.fm

P. Melo
Centre for Business and Economics Research, Faculty of Economics, INESC Coimbra, University of Coimbra, Av. Dr. Dias da Silva 165, 3004-512 Coimbra, Portugal
e-mail: pmelo@fe.uc.pt

V. Raposo
Centre for Business and Economics Research (CeBER), Centre of Health Studies and Research of the University of Coimbra, Coimbra, Portugal
e-mail: vraposo@fe.uc.pt

© The Author(s) 2022
D. Kriksciuniene and V. Sakalauskas (eds.), *Intelligent Systems for Sustainable Person-Centered Healthcare*, Intelligent Systems Reference Library 205,
https://doi.org/10.1007/978-3-030-79353-1_6

91

of evaluation for healthcare; the third one follows with illustrations of patient-centred and person-centred methods of evaluation; and the last part moves forward with a reflection on intangibles and a proposition for a method of observation.

Keywords Efficiency evaluation · Methods · Person-centered · Value

6.1   Introduction

Nowadays, evaluation appears as a necessity. Yet, there is no consensus on what counts or what does not count as evaluation criteria. "What cannot be counted does not count" is a commonly heard sentence. But, is what is counted what matters the most? Evaluation traditionally means *conferring value to a process, person or organization.* The content of this "value" depends on underlying assumptions about what is valuable and the tool or technique used to evaluate.

In the healthcare sector, the cost-control rhetoric became prominent with the reference of decision-making processes. Health economics and managers use measurement tools to determine costs, benefits and what it is worth doing in healthcare organizations. Prices are supposed to reflect the value produced by one hospital or one clinic. Efficiency is understood as the result of these outputs. The purpose of health economic evaluation is to identify and sustain efficiency within the healthcare system because it influences decision-making processes and policy design [80]. Classic health economics balances ins and outs according to types of costs and benefits [18]: direct (resources use), indirect (patient's time), and intangible (patient's condition, pain). On a more systemic level, one may consider outcomes as a basis for evaluating a care process [71, 72]. However, we can all agree on the fact that mere cost-cutting is detrimental to care, there is still debate on the nature of outcomes we should investigate. For Porter, "what is not measured can't be managed" [45]. This kind of formulation does not address chronic diseases: How to measure, not only the costs, but the added value of care for a person going through a long-term protocol? How can we achieve this when the healthcare sector is dominated by evidence-based medicine, a medicine built on measurable proofs and outcomes [8]?

Measuring tools are also not neutral. They do not measure pre-existing performance. They construct the very notion of performance. Performance, in turn, can be approached in a variety of ways: economic, organizational, social…the notion itself puts "competing values" into play [75]. Economic evaluation confers an undisputed value to care by breaking it down into technical components and matching them with single costs. Efficiency can be classically assessed through a number of indicators (readmission rate, mortality rate, morbidity rate, number of visits…). From this standpoint, the intangible part of care, relationships and acceptance, is left aside. Any numerical indicator hardly takes into account the relational component of care, which

Faculty of Economics, University of Coimbra, Av. Dr. Dias da Silva 165, 3004-512 Coimbra, Portugal can be much more complex to grasp [53]. Empathy, sensitivity to needs or relationship building are crucial but evanescent when it comes to talking numbers. What is valued by some patients can make others uncomfortable. For instance, talking about oneself is frowned upon in some cultures. Tak et al. [85] even found that if patients do not have all the information to evaluate the quality of care, this notion remains their main criteria of satisfaction, more so than explanation or listening skills. The value of care also depends on the values of the patient. How can we integrate their perspectives into evaluation of care given this ambiguity and heterogeneity?

The chapter is structured as follows. In the first part, an original cross-country analysis of healthcare systems performance, based on a dual efficiency/effectiveness model, sets the stage in the European context. The second part goes back to the fundamental principles of methods of evaluation in healthcare. Delivering value-added healthcare services is not insignificant in the choices of methods, indicators, factors and underlying concepts. The third part illustrates key advancements in evaluation methods such as patient and person-centered settings with results of effective interventions that are rooted in different health economics paradigms. The final section aims at moving forward and proposing a different method rooted in management research qualitative methods to take into account the intangibles of healthcare in evaluation paradigms.

6.2   General Overview of Healthcare Systems: What is at Stake. Cross-Country Evaluation of the Performance of Healthcare Systems in Europe

Since the early 2000s the European healthcare systems have been facing several challenges [66] including: (a) the increasing costs of healthcare; (b) the ageing of population associated with the rise of chronic diseases and the growing demand for healthcare; (c) the lack of equity in accessing healthcare services; (d) an uneven distribution of healthcare professionals and infrastructure assets across regions. However, the budget restrictions in the public sector which have occurred in the last decades, before the COVID-19 pandemic, have limited financial resources, jeopardizing the sustainability of national healthcare systems and the possibility to deliver high quality health care service and provide universal access. Hence, the need to deliver value-added healthcare services focusing on resource and cost efficiency and increasing health quality has become an important goal in the changing landscape of healthcare management in Europe. Indeed, healthcare consumes a large percentage of national budgets, and not all countries are able to get an acceptable value for their investment money. According to data available from the World Bank database [95], in 2018 Norway, Switzerland and the United States were the biggest spenders in healthcare in the world, respectively having a health expenditure per capita of (current US$) $8,239 (10.1% of GDP), $9,871 (11.9% of GDP), and $10,624,403 (16.9% of GDP). However, in the same year the healthcare systems in other countries were achieving similar or even better results by spending far less. For instance, expenditure per capita was $2,989 (8.7% of GDP) for Italy, $3,323 (7.5% of GDP) for Israel, $2,754 (9.0% of GDP) for Malta, and $2,824 (4.5% of GDP) for Singapore respectively. Life expectancy in all these countries is between 82 and 84 years as in Norway and Switzerland, higher than in the United States, in which it is 79 years.

Notwithstanding some important factors like lifestyles, diet, pollution, etc. which affect life expectancy, the way healthcare services are delivered to the general population and the way that healthcare management systems are designed and implemented play a critical part. Both costs and performance of the national healthcare systems can be explained in terms of their design, organization, implementation and management. National healthcare systems differ between European countries because cultural norms, market regulations, policies, and history have shaped each of them. However, although there are differences in terms of infrastructure endowment, patient population size, fund allocation, and management settings, they face similar challenges and have common goals. Thus, assessing and comparing the performance of several national health care systems provides an opportunity for policy makers to determine how well the country healthcare system is performing relative to its international peers, understand how it works in order to identify good and bad practices, and finally find more effective approaches to achieve sustainability and better quality [63]. Identifying performance indicators and developing measurement frameworks have become an important concern of policy makers and scholars [1]. Both international agencies and academic scholars have proposed various sets of metrics, benchmarking tools, assessment guidelines, and performance evaluation techniques to help healthcare policy makers to monitor and evaluate the performance of the national health systems and conduct benchmarking studies both at the national and international level [97]. However, performance evaluation and benchmarking models are still far from being developed and capable to provide useful results in healthcare planning. Additionally, academic and industry literature reports evidence of diffused inefficiency in healthcare management in Europe, contributing to increases in health expenditure in the last decade [41, 65]. Furthermore, empirical evidence [56] indicates that high level of efficiency cannot be achieved without reducing quality or effectiveness of healthcare service provision due to potential trade-off between them. Thus, developing a performance framework and metrics that focus on the process that transforms resources into healthcare outcomes still remains an important topic for researchers and public policy makers.

The literature for the last two decades has found a huge number of publications focusing on the measurement of efficiency in the healthcare sector. However, there are relatively few studies that evaluated and compared efficiencies of healthcare systems at the country level [91]. Since the seminal study by the World Health Organization [96] on the efficiency of the health systems in 191 countries around the world, there has been a growing scholarly interest to develop performance metrics to assess and compare the national healthcare systems and to investigate determinants of either unacceptable or outstanding performance.

Certain studies are based on the utilization of individual performance indicators [33]. Such performance indicators are generally derived from publicly available data [97]. Sometimes, individual performance indicators are combined together to obtain homogeneous groups of countries whose healthcare systems achieve comparable performance measurements along multiple dimensions [86]. Some studies rank country healthcare systems and identify determinants of efficiency by implementing various econometric models [3, 6, 25, 96].

Most studies use either parametric and non-parametric analytical techniques such as the Stochastic Frontier Analysis (SFA) or the Data Envelopment Analysis (DEA), in which the healthcare systems are modelled as decision-making units [28, 40]. It seems the DEA is preferable to evaluate efficiency due to a high number of advantages: it gives an opportunity to include in a model several inputs and outputs that allows estimating efficiency without calculation of a sole parameter of input or output; absence of necessity to choose the functional form of production function; it allows to analyse the efficiency in cases when it is difficult enough formally to explain relation between numerous inputs and outputs of a system; it enables to estimate the contribution of each of inputs to overall efficiency (or inefficiency) of the decision-making units and to estimate a level of inefficiency of each input; and besides an estimation of technical efficiency, it enables to estimate other kinds of efficiency, e.g. economic efficiency [34]. Hence it is apparently more commonly used to evaluate healthcare efficiency of healthcare. Because of this, Bhat [7] has adopted DEA to assess the influence of specific financial and institutional arrangements on national healthcare system efficiency in a sample containing 24 OECD countries. It was found that countries having public-contract and public-integrated based healthcare systems are more efficient than those having public-reimbursement based systems. Afonso and St Aubyn [2] performed two-stage DEA, estimating a semi-parametric model of the healthcare system in 30 OECD countries in 1995 and 2003. They computed conventional and bootstrapped efficiencies in the first stage and corrected these values in the second stage by considering the influence of non-discretionary variables such as GDP per head, education level, and health behaviour using Tobit regression. Results show that a large amount of inefficiency is related to variables that are beyond the government control. Gonzalez et al. [35] measured the technical and value efficiency of the health systems in 165 countries using data for 2004. They used data on healthy life expectancy and disability adjusted life years as health outcomes, and the amount of expenditure on health and education as inputs to the healthcare system. Findings reveal that high-income OECD countries have the highest efficiency indexes. Likewise, Varabyova and Schreyögg [91] compared the efficiency of the healthcare systems using an unbalanced panel data from OECD countries between 2000 and 2009. In particular, they used different model specifications performing two-step DEA and one-stage SFA and assessed internal and external validity of findings by means of the Spearman rank correlations. Their study shows that countries having higher healthcare expenditure per capita have on average a more efficient healthcare sector, while countries with higher income inequality have less efficient healthcare.

Hadad et al. [39] compared the healthcare system efficiency of 31 OECD countries utilizing various efficiency conceptualizations (conventional efficiency, super-efficiency, cross-efficiency) and two model specifications, one including inputs that are under management control and another incorporating inputs that are beyond management control. The study provided ambiguous results. Kim and Kang [48] estimated the efficiency of the healthcare systems in a sample of 170 countries performing bootstrapped DEA. The sample was organized into four groups to obtain homogeneous sub-samples with respect to income. Scholars found that average efficiency in the high-income sub-sample was relatively high, but only a small number of the countries are able to manage their healthcare systems efficiently. De Cos and Moral-Benito [12] investigated the most important determinants of healthcare efficiency across 29 OECD countries estimating alternative measurements of efficiency performing DEA and SFA from 1997 to 2009. Their study provides empirical evidence that there are significant differences among countries with respect to the level of efficiency in healthcare services provision. Furthermore, there is a positive correlation between the implementation of policies aimed at increasing price regulation and the efficiency of the national healthcare system. Frogner et al. [27] measured healthcare efficiencies of a sample including 25 OECD countries between 1990 and 2010 using publicly available data. Three econometric approaches were adopted, i.e. country fixed effects, country and time fixed effect models, and SFA including a combination of control variables reflecting healthcare resources, behaviours, and economic end environmental contexts. The study shows that rankings are not robust due to different statistical approaches. The study by Kim et al. [49] estimated productivity changes in the healthcare systems of 30 national healthcare systems during 2002–2012. Scholars calculated the bootstrapped Malmquist index to analyse changes in productivity, efficiency and technology. They found that recent policy reforms in OECD have stimulated productivity growth for most countries (Fig. 6.1).

This literature review shows that scholars mostly focused on the measurement of one single index of healthcare system performance, i.e. the efficiency calculated as a ratio of a measure of the quality of life to the amount of health resource used. No effectiveness estimates are generally used in the analyses. This shortcoming has been eliminated by Lo Storto and Goncharuk [54], which suggested dual efficiency/effectiveness model for cross-country evaluating the performance of healthcare systems. Lo Storto-Goncharuk's model uses DEA for presenting and comparing efficiency and effectiveness scores for every national healthcare system

Fig. 6.1 Cumulative productivity growth, 2002–2012 (2002 = 100). *Source* Kim et al. [49]

Table 6.1 Inputs and outputs

Code	Type	Description	Measuring
I1	Input	Medical doctors (practicing)	No. of units
I2	Input	Nurses, midwives, healthcare assistants (practicing)	No. of units
I3	Input	Available beds in hospitals	No. of units
O1	Output (bad)	Ratio of infant mortality (less than 1 year) to population	Percentage
O2	Output (good)	Healthy life years in absolute value at birth (both males and females)	No. of years
O3	Output (good)	Life expectancy in absolute value at birth (both males and females)	No. of years
O4	Output (good)	population	No. of units

Table 6.2 DEA models implemented

	Index	Inputs	Outputs	Orientation
Model 1	Efficiency of the healthcare system	I1, I2, I3	O4	Input
Model 2	Effectiveness of the healthcare system	O1	O2, O3	Output

in two-dimensional space. Since this model requires only publicly available statistics (Table 6.1), it allows evaluation and comparison of the effectiveness and efficiency of healthcare systems in various countries, for example in European countries.

As Lo Storto and Goncharuk [54] have suggested, benchmarking analysis was used to implement two DEA models as illustrated in Table 6.2. For both models, constant returns to scale have been assumed.

Applying these two models for 32 European countries for 2011–2014 period, the authors found the most efficient healthcare systems in Europe (Irish, Polish and Portugal systems) and the most inefficient (Lithuania, Norway, Switzerland, Germany and Austria). Effectiveness proved to be more dynamic than efficiency. Between 2011 and 2014, two countries made fantastic breakthroughs in effectiveness of healthcare: Slovenia by over 100% and Cyprus by 200%. So, at the end of 2014 these countries had the relatively highest healthy life years and life expectation together with the lowest infant mortality.

Comparing the efficiency and effectiveness scores, Lo Storto and Goncharuk [54] identified a group of countries with the least successful healthcare systems. It included Romania, Ukraine and Bulgaria. It was concluded that these countries need to implement healthcare reforms aimed at reducing resource intensity and increasing the quality of medical services.

Given the somewhat outdated results of the study by Lo Storto and Goncharuk [54], we decided to update them and figure out whether there have been significant changes in the levels of relative performance of national health systems of the same 32 European countries.

In addition, we decided to refine the output O4 in model 1 (efficiency of the healthcare system), since we believe that the entire population is not a completely appropriate output of the healthcare system work. Hence, we replaced it with the number of people with good or very good perceived health. These statistics with the data on three inputs of model 1 we got from Eurostat and State Statistics Service of Ukraine for 2017. The main statistics for 2011, 2014, and 2017 are described in Table 6.3.

The model1-cor. means the model for evaluating the efficiency of healthcare with changed output (number of people with good or very good perceived health). However, the model1 means the same model as in Lo Storto and Goncharuk [54].

The results of cross-country evaluations on the efficiency of 32 European healthcare systems for 2017 using model1 and model1-cor. can be seen at Fig. 6.2.

Our correction of the model 1 gave higher differences for efficiency scores of healthcare systems. Apparently, a noticeable lower efficiency scores from corrected model 1 for such countries as Estonia, Latvia, Lithuania, Portugal, and Ukraine reflect a lower percentage of people there with good or very good perceived health in comparison to the other European countries. In addition, according to the scores

Table 6.3 Main statistics relative to DEA models

	2011		2014		2017	
	Model1	Model2	Model1	Model2	Model1	Model1-cor
Mean	0.643	0.324	0.660	0.439	0.790	0.717
St.dev	0.154	0.160	0.157	0.181	0.133	0.032
Max	1.000	1.000	1.000	1.000	1.000	1.000
Min	0.417	0.114	0.459	0.167	0.522	0.331

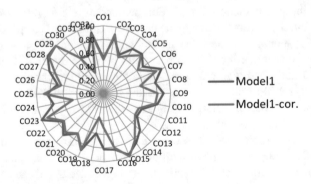

Fig. 6.2 Efficiency scores for European countries during 2011–2017. Notes: Austria (CO1), Belgium (CO2), Bulgaria (CO3), Croatia (CO4), Cyprus (CO5), Czech Republic (CO6), Denmark (CO7), Estonia (CO8), Finland (CO9), France (CO10), Germany (CO11), Greece (CO12), Hungary (CO13), Iceland (CO14), Ireland (CO15), Italy (CO16), Latvia (CO17), Lithuania (CO18), Luxemburg (CO19), Malta (CO20), Netherlands (CO21), Norway (CO22), Poland (CO23), Portugal (CO24), Romania (CO25), Slovakia (CO26), Slovenia (CO27), Spain (CO28), Sweden (CO29), Switzerland (CO30), Ukraine (C31), United Kingdom (CO32)

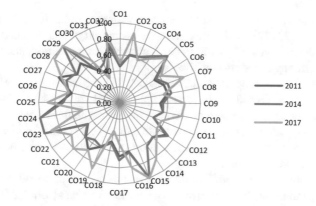

Fig. 6.3 Efficiency scores during 2011–2017 from two models of evaluation. *Notes* Austria (CO1), Belgium (CO2), Bulgaria (CO3), Croatia (CO4), Cyprus (CO5), Czech Republic (CO6), Denmark (CO7), Estonia (CO8), Finland (CO9), France (CO10), Germany (CO11), Greece (CO12), Hungary (CO13), Iceland (CO14), Ireland (CO15), Italy (CO16), Latvia (CO17), Lithuania (CO18), Luxemburg (CO19), Malta (CO20), Netherlands (CO21), Norway (CO22), Poland (CO23), Portugal (CO24), Romania (CO25), Slovakia (CO26), Slovenia (CO27), Spain (CO28), Sweden (CO29), Switzerland (CO30), Ukraine (C31), United Kingdom (CO32).

from two models, only Sweden and Ireland had efficient healthcare systems in 2017 (Fig. 6.3).

6.3   Underlying Concepts and Definitions of Evaluation Methods for Healthcare

However, to properly assess the healthcare evaluation trade-offs, the used evaluation framework characteristics must be considered. Therefore, we will now present a short digression over the main current economic evaluation perspectives proposed currently.

Value-based healthcare (VBHC), Value for Money (VfM) and economic evaluation helped to change the paradigm of healthcare systems and have been central to health policy decisions, accountability, healthcare delivery and healthcare systems [52, 59]. In the next sections, we look for these main concepts focusing on their definition, importance, advantages and limitations.

6.3.1   Value-Based Healthcare

VBHC is a healthcare delivery model in which providers, including hospitals and physicians, are paid based on patient health outcomes [72]. The value in VBHC is derived from measuring health outcomes that matter to patients against the cost of delivering them [72, 73]. In this model, the relevant unit of analysis is delivered to a patient over the full cycle of care for a particular medical condition, such as diabetes, breast cancer or any other chronic disease.

According to several reports and authors [24, 87], the benefits of a VBHC system extends to patients, providers, payers, suppliers, and society as a whole: patients spend less money to achieve better outcomes; providers achieve greater patient satisfaction and better care efficiencies; payers have strong cost control and reduced risks; suppliers can align prices with patient outcomes; society becomes healthier while reducing healthcare spending.

Evidence shows that various health care systems across the world have embraced the VBHC agenda for different reasons over the last 15 years [59] and it has become a guiding principle in the quest for high-quality health care with acceptable costs [37]. Several reports have evaluated the implementation status of VBHC across the world [81, 88] and recently the European Institute of Innovation & Technology (EIT) Health published a handbook on how to adopt VBHC initiatives [20]. Mjåset et al. [59], mention that although no country has fully implemented the VBHC agenda, it seems apparent that different theoretical framework elements function better in some healthcare systems than others.[1]

However, not everyone is convinced that the VBHC guide is the appropriate way forward. According to Nilsson et al. [62], value for patients was experienced as the fundamental drive for implementing VBHC, but there are multiple understandings in parallel of what value for patients means. In the same line, Pendleton [68], using the results from a survey conducted in the United States, states that different stakeholders have no common definition of value and do not agree on its composition. He also says that value seems to have become a buzzword with its meaning often unclear and shifting, depending on who is setting the agenda. Groenewoud et al. [37] also argue that current literature lacks substantial ethical evaluation of VBHC and that a single-minded focus on VBHC may cause serious infringements on medical ethical principles.

Groenewoud et al.'s [37] arguments focus on several points: (lack of) evidence of VBHC effectiveness on more efficient clinical pathways, due to scarcity of transparency, cost awareness and relevant outcomes; (lack of) evidence of translatability of VBHC concepts from business strategy to health care; (lack of) match between business ethics and healthcare values; (and lack of) a common ontology, since the concept of values from ethics and philosophy is different of the VBHC approach (outcomes divided by costs). The main infringements identified are related to neglecting four medical ethical principles: it tends to neglect patients' personal values; ignores the intrinsic value of the caring act; disproportionately replaces trust in professionals with accountability, and undermines solidarity.

---

[1] They based this conclusion after comparing the various health care funding schemes (more private or more public) and the six VBHC elements proposed by Porter and Lee [74]: care organized around medical conditions, outcome and cost measurement for every patient, value-based reimbursement for all the care cycles, regional systems integration, the geography of care with centers of excellence, and information technology supporting VBHC.

Besides these questions, there are also other problems related to the full implementation of VBHC, namely to calculate values associated with the determination of the health outcomes that matter to patients (numerator) and the costs of delivering the outcomes (denominator). These problems arise when we consider the complexity of healthcare providers, the heterogeneity of management processes, and the different services provided with cost systems that do not directly calculate the involved costs. Some of these problems are also faced by VfM and (generic) economic evaluation.

6.3.2 Value for Money and Economic Evaluation

VfM and economic evaluation have been central to the healthcare systems agenda in questions about health policy decisions, accountability, and care delivery [52, 64, 79]. The main point is that resources (people, time, facilities, equipment, and knowledge) are scarce and choices must be made avoiding traditional heuristics like "do what we did last time", "follow gut feelings", or "educated guesses" [19].

In the context of managing constrained healthcare budgets and safeguarding equity, access and choice, governments face the challenge to strategically manage scarce resources by investing in services that provide the best health outcomes [5, 52, 64, 82]. The economic evaluation provides a common framework that helps to identify the relevant alternatives, facilitates the integration of different perspectives and viewpoints (patient, institutional, target groups, and other stakeholders), reduces subjectivity by raising quantification over informal assessment, and increases the explicitness and accountability in decision-making [19, 31]. VfM and economic evaluation also reinforce accountability by ensuring that taxpayers' money and other funding instruments are spent wisely, and assuring healthcare users and other stakeholders that their claims and interests on the health system are being treated fairly and consistently [82, 79, 26, 43].

Drummond et al. [19] define economic evaluation as the comparative analysis of alternative courses of action in terms of both their costs and consequences. In any economic evaluation, the main tasks are to identify, measure, value, and compare the costs and consequences of the different alternatives being considered. However, given the nature of the consequences, especially in the healthcare field, considering the options being examined may differ considerably.

VfM includes the four E's in its assessments [26, 42]: savings (minimizing the cost of inputs, while bearing in mind quality); efficiency (achieving the best rate of conversion of inputs into outputs, while taking in mind quality); effectiveness (achieving the best possible result for the level of investment, while maintaining in mind equity); and equity (ensuring that benefits are distributed fairly).

For Smith [79], VfM can be examined from several perspectives: the economic perspective, concerned with which physical inputs are purchased; the extent to which the chosen inputs are combined in an optimal mix; the technical efficiency with which physical inputs are converted into physical outputs; the allocative efficiency of the system's chosen outputs; and the quality of the care provided (its effectiveness). To this author, the two fundamental managerial tasks are purchasing decisions (allocative efficiency) and performance assessment (technical efficiency).

Fleming [26] identifies six main methods that can be used to assess VfM: Cost-Effectiveness Analysis (CE analysis), Cost-Utility Analysis (CU analysis), Cost–Benefit Analysis, Social Return on Investment (SROI), Rank correlation of cost vs impact, and Basic Efficiency Resource Analysis (BER analysis). Smith [79] claims that in the VfM field, in parallel to the piecemeal analysis of individual performance measures, most of the research is under the label of productivity analysis, using econometric methods, such as stochastic frontier analysis (SFA); or descriptive methods known as data envelopment analysis (DEA).

The primary purpose of economic evaluation is to inform decisions, so it deals as mentioned before, with both inputs and outputs (costs and consequences) of alternative courses of action and is concerned with choices. The main types of economic evaluation studies are cost analysis (without identification or measurement of consequences), cost-effectiveness analysis, cost-utility analysis, and cost-benefit analysis [19].

6.3.3 Measuring Costs and Consequences

Most of the considered concepts presented—VBHC, VfM and economic evaluation—are concerned with choices when comparing costs and consequences (economic, clinic and humanistic[2] outcomes). In the next sub-sections, we focus on the main issues related to costs and consequences (different types, difficulties of measurement and possible sources that can help in the selection of tools).

Costs

Focused on a cost-effectiveness analysis, Gold et al. [31] identifies costs related to changes in the use of healthcare resources, changes in the use of non-healthcare resources, changes in informal caregiver time and changes in the use of patient time (for treatment). The same author identifies different types of costs:

- Direct health care costs—all types of resource use, including the consumption of professional, family, volunteer, or patient time and costs of tests, drugs, supplies, healthcare personnel, and medical facilities.
- Non-direct health care costs—include additional costs related to the interventions like those for childcare (for a parent attending a treatment), the increase of costs required by a dietary prescription and the costs of transportation to and from health facilities; it also includes the time family, or volunteers spend providing home care.

---

[2] In the ECHO (Economic, Clinical, and Humanistic Outcomes) model, medical care outcomes can be classified along 3 dimensions: clinical, economic, and humanistic. «Humanistic outcomes included measures of patient satisfaction and patients' quality of life», see Cheng et al. [14].

- Patient time costs include the time a person spends seeking care or participating in or undergoing intervention or treatment. Relevant time costs include travel and waiting time as well the time receiving treatment.

On the other hand, in a broader perspective of economic evaluation involving costs and different types of analysis, Drummond et al. [19], identifies health sector costs, other sector costs, patient/family costs, and productivity losses:

- Health sector costs can be variable (such as the time of health professionals or supplies) and fixed or overhead costs (such as light, heat, rent, or capital costs).
- The other sector costs refer to consumed resources from other public agencies or to the voluntary sector.
- Person/family costs refer to any out-of-pocket expenses incurred by patients or family members and the value of any resources that they contribute to the treatment process.
- Productivity costs include the costs associated with lost or impaired ability to work or to engage in leisure activities due to morbidity and lost economic productivity due to death.

This way, several authors identify categories of direct costs, indirect costs and intangible costs. Direct costs associated with providing the health service (fixed, variable, and non-medical expenses) are the easiest to calculate. Indirect costs related to decreased productivity due to the disease or treatment in the patient and his family are difficult to compute. Intangible costs (such as anxiety, pain or suffering with an illness) are extremely difficult or even impossible to determine.[3] These problems with the cost measurement are common to VBHC, VfM and economic evaluation.

One example of this problem in the VBHC is the determination of hospitals costs. Hospitals are very complex organizations [17, 32], with quite distinct management processes joining the worlds of care, cure, control and community [29, 30], with different types of services, clinical pathways, treatments and decisions with cost systems more oriented to the disease than to the patient. According to Kaplan et al. [45], the existing cost systems in healthcare prevent clinician-driven cost reduction and process improvement initiatives, and time-driven activity-based costing (TDABC) is one tool with significant potential to fill this gap. The same author argues that these systems rely on inaccurate and arbitrary cost allocations and provide little transparency to guide first-line care providers attempts to understand and change the proper drivers of their costs.

This approach has several advantages identified in the literature: more accurate cost estimates [13], efficiency in allocating costs to the cost object [47], better use of resources, activities and processes, increasing the capacity used and eliminating those that do not add value [22, 46, 98], more accurate allocation of indirect expenses to the cost object [46], process optimization, trying to reduce time consumed by some activities [36], and the best benchmarking model [78]. However, despite all these advantages, the possible inaccuracy of time estimates [16, 36] and the time

---

[3] Section 4 will propose a method to grasp intangibles in healthcare.

needed to determine time estimates [78] hinder its implementation in healthcare and applications of TDABC to healthcare have been limited [46].

Consequences

The benefits of a VBHC system extend to patients, providers, payers, suppliers, and society as a whole [24, 87]. Many treatments offer broader social and economic benefits to patients, families and society [79]. The responsiveness to patients' needs, addressing inequalities, and broader economic objectives are the leading healthcare goals of healthcare systems. Some authors also focus on economic outcomes and their interrelationships with the clinical and humanistic outcomes [14].

The literature on economic evaluations identifies several types of outcome measures like clinical outcomes, quality of life measures, and generic health gain measures like Quality-adjusted life years (QALYs), the Disability-Adjusted Life-Year (DALY), SF-36, EQ5D, and SF6D [19, 31, 82).

Clinical outcomes are the most common health outcome category to be considered in clinical trials and observational studies. Humanistic outcomes are outcomes based on a patient's perspective (e.g., patient-reported scales that indicate pain level, degree of functioning). In this category, there are health-related quality of life (HRQoL) and the range of measures collectively described as patient-reported outcomes (PRO), which include measures of HRQoL3.

Patient-reported outcomes (PROs), or patient-reported outcomes measures (PROMs), are information provided by the patient about their symptoms, quality of life, adherence, or overall satisfaction [55, 92]. PROs refer to patient ratings about several outcomes, including health status, health-related quality of life, symptoms, functioning, satisfaction with care, and treatment satisfaction. The patient can also report about their health behaviours, including adherence and well-being habits. Data is collected by generic and disease-specific validated tools related to the quality of life (e.g. EQ-5D, AqoL), symptoms (e.g. NPRS for pain, FSS for fatigue), distress (e.g. K10 or PHQ2 for depression, GAD7 for anxiety), functional ability (e.g. WHODAS 2.0, ODI), self-reported health status (e.g. SF-36), or self-efficacy (e.g. GSE).

Patient-reported experience measures (PREMs) are tools and instruments that report patient satisfaction scores with health service [92]. They are generic tools that are often used to capture the overall patient experience of health care. PREMs are often used on the broader population and in non-specific settings such as an outpatient department. Patient experience tools for example may be used to monitor patient feedback and focus on the general experience related with time spent waiting, the access to and ability to navigate services, the involvement (consumer and carer) in decision-making, the knowledge of care plan and pathways, the quality of communication, the support needed to manage a long-term condition, if they would recommend the service to family and friends, etc. They are a reliable measure of how well a hospital or other health unities provide good quality service from a patient perspective.

According to Lavallee et al. [51], the time devoted to collecting PROs and PREMs is a time investment that can benefit the person receiving care and the organization that can allocate resources more optimally. Assessing the severity of symptoms, informing treatment decisions, tracking outcomes, prioritize patient-provider discussions, monitoring general health and well-being, and connecting providers to patient-generated health data are different ways of creating value.

There are several organizations where it is possible to find different tools for the purposes mentioned above. The International Consortium for Health Outcomes Measurement (ICHOM) collaborates with patients and healthcare professionals to define and measure patient-reported outcomes to improve care quality and value. In the ICHOM website[4] several standardized outcome measurement tools are presented, as well as time points and risk adjustment factors for a given condition. The Patient-Reported Outcomes Measurement Information System (PROMIS) website[5] includes over 300 measures of physical, mental, and social health for use with the general population and individuals living with chronic conditions. The Outcome Measures in Rheumatology (OMERACT)[6] is an independent initiative of international stake-holders interested in outcome measurement. The Consensus-based Standards for the selection of health Measurement (COSMIN)[7] aims to improve the selection of outcome measurement instruments both in research and in clinical practice by developing methodology and practical tools for selecting the most suitable outcome measurement instrument.

A recent categorization of data is patient-reported information (PRI), proposed by Baldwin et al. [4]. According to those authors, PRIs upgrades the PRO tool rein-forcing the patient perspective. This new perspective is related to social networking, enabling patients to publish and receive communications quickly. Many stakeholders, including patients, are using social media to find new ways to make sense of diseases, to find and discuss treatments, and to give support to patients and their caregivers. According to Schlesinger et al. [77], PRI pinpoints the limits of tradi-tional measurement techniques to incorporate narrative components into the evalua-tion and can be used to improve clinical practice. Those authors identify four forms of PRIs: (1) patient-reported outcomes measuring self-assessed physical and mental well-being, (2) surveys of patient experience with clinicians and staff, (3) narra-tive accounts describing encounters with clinicians in patients own words, and (4) complaints/grievances signalling patients distress when treatment or outcomes fall short of expectations.

6.4   Patient-Centred Versus Person-Centred Evaluation Methods: Illustrations

A health economic evaluation can be conducted from one of the six perspec-tives (public-health, health care system, healthcare payers', institutional and/or patients' perspective). Health economic analysis is almost performed as an aid to

---

[4] https://www.ichom.org.

[5] https://www.healthmeasures.net.

[6] https://omeract.org.

[7] https://www.cosmin.nl.

the medical decisions of healthcare facilities and healthcare systems, leaving the patient perspective out of the equation.

With the recent shift from patient-centered care to person-centered care in both of these approaches, the role of the patient in treatment decisions plays an important role in health policy. Encouraging patients to participate in decision making is not easy to do, but it is becoming a norm among growing evidence that health outcomes are often observed from the patients' perspective in terms of health quality, patient preference, and/or part of patients' health care costs.

In healthcare organizations that are patient-centered and person-centered oriented, the primary economics benefits concern lower medical costs compared to usual care settings. Health economic evaluation, patient-cantered care, and person-centered care are difficult to directly compare because the available studies are different in terms of methods used, type of costs and outcomes measured, the patient population of interest, and various types of interventions. The studies are also conducted in different health systems with specific socioeconomic environments and cultures.

Person-centerd care and like healthcare approaches have shown beneficial effects and lower costs [70]. Extending person-centred care in healthcare practice demands more cognition about the effects and the cost-effectiveness of person-centred care. Most studies have shown that person-centred care is cost-effective compared to usual care [70].

This subchapter presents illustrative results of some effective intervention ("3D", Dementia Care Mapping and Palliative Advanced Home Care and Heart Failure Care) from different perspectives of health economics.

6.4.1 "3D" (Dimensions of Health, Depression and Drugs) Intervention

The "3D" intervention was developed to address the issues associated with managing patients with multimorbidity in primary care in the UK [89].[8] The number of patients living with multiple chronic health conditions (multimorbidity) is indeed increasing as the population is ageing [60]. The prevalence of multimorbidity is approximately 98% for older adults. As the elderly population grows, a complex cost-effectiveness intervention is needed at different levels of the healthcare system.

There is no evidence that a comprehensive multimorbidity care programme has reduced healthcare costs or primary care visits. There were many ways to organise patient care to take into account multimorbidity, but evidence of effectiveness and recommended strategies is limited.

The "3D" intervention evaluated a patient-centred care approach for patients with three or more long-term conditions. The approach included improvement of the conti-nuity of care and regular holistic review ("3D": nurse, pharmacist, and general prac-titioner (GP) in general practice (GP) surgeries. The intervention aimed at reducing

---

[8] "3D" intervention model is well documented by Thorn et al. [89], whom we rely on in this section.

the burden on the patients in accessing healthcare and increasing patient participation in decision-making about their care. Also, nurse specialists usually carry out a review of chronic conditions for particular conditions in primary care.

Quality adjusted life years (QALYs), as part of outcome measurements, uses the EQ-5D-5L[9] 15 months after randomisation. This trial used cost-utility analysis conducted from the perspective of the NHS and personal social services [61].

The primary analysis showed that the participants in the intervention group gained a mean of 0.007 (95% CI: −0.009 to 0.023) additional QALYs over 15 months compared with participants in the usual care group [89]. From the NHS/PPS perspective the total cost per patient was £126 (95% CI: −£739 to £991) higher in the intervention group than in the usual care group (Ibid). A cost-effective analysis showed that the ICER[10] was £18.449 and the net monetary benefit in terms of societal willingness to pay the value of £20.000 was £10 (95% CI: −£956 to £977). The sensitive cost-effectiveness of the "3D" approach has showed that this approach was associated with lower costs and better outcomes.

The beneficial effect of this intervention on patient care experience is more person-centred, but modifications that support better implementation are needed to improve the intervention's effectiveness.

6.4.2   Cost-Effectiveness and Cost Dementia Care Mapping[11]

It is estimated that more than 35 million people worldwide have dementia and expect their number to grow. The course and outcomes of dementia vary from patient to patient, but the condition usually has significant effects on quality of life, as a result of one or more behaviours. The following behaviours are described as a challenge to support (BSC): agitation, aggression, restlessness, hallucination, delusions, depressions, anxiety, and apathy.

Dementia Care Mapping (DCM) is a widely used intervention at the home care level to observe patients with dementia. This intervention aims were to improve individual person-centred care, the quality of healthcare and health outcomes for residents. It has been widely used to cure dementia for almost twenty years. Despite widely used evidence of cost-effectiveness, randomised and non-randomised interventions are mixed. Only two studies report on economic evaluation of the intervention and none on a cost-utility analysis.

The DCM-EPIC[12] is a pragmatically randomised controlled trial aimed at evaluating clinical and cost-effectiveness, a controlled trial of usual care plus DCM

---

[9] See: https://euroqol.org/eq-5d-instruments/eq-5d-5l-about/EQ-5D-5L measures health-related quality of life in cost-effectiveness analysis.

[10] Incremental cost-effectiveness ratio.

[11] See: Surr et al. [84].

[12] Dementia Care Mapping™ to enable person-centred care for people with dementia and their carers (DCM-EPIC).

(intervention group) and compared to usual care (control group). DCM has been implemented using standard procedures and following the most common UK model of staff-led use implementation [84]. Two staff members ("mappers") from the intervention home care attended four days of standard training in DCM [84]. Cost-effectiveness analysis measured incremental costs, CMAI[13] and QALYs for residents.

This DCM trial results were not found to be effective versus control on the primary or secondary outcomes, nor was it cost-effective (Ibid.). The cost for unit improvement in the CMAI is higher than other recent evaluation of interventions that include training of staff in person-centred care or communication skills with or without behaviour management training (Ibid.). Also, the cost per QALYs was higher than the upper bound of the threshold over which treatments are least likely to be funded in England. (Ibid.).

A complex system-level intervention like this one, which used staff-led implementation, may not provide a real implication intervention without applying other implementation models to optimise the intervention. Barriers and facilitators on DCM implementation were at the mapper and care home level. The barriers at these levels include the lack of mapper time, skills, and confidence to implement DCM, lack of resources, and management support (Ibid.).

Another study, which used the DCM model, also did not find the method effective versus the control group and suggested that future research should investigate value for money as an alternative strategy to prevent and support behaviour symptoms in people living with dementia in care homes.

6.4.3   Cost-effectiveness Palliative Advanced Home Care and Heart Failure Care (PREFER) Intervention[14]

Chronic heart failure (CHF) is a significant public health issue worldwide. In developed countries, approximately 1–2% of the population has CHF, and the prevalence is rising in people over 70 years old [10].

The randomised controlled study confirmed that Palliative Advanced Home Care and Heart Failure Care (PREFER) improve patients' quality of life and reduce health care costs due to reduced number of hospitalisation days and reduced number of hospitalisations [9]. When the person-centred care was fully implemented to patients with CHF, the length of hospitalisation was reduced [21]. This randomised control study's primary aim was to assess the cost-effectiveness of the PREFER intervention compared to standard care for patients with heart failure [76].

---

[13] Cohen-Mansfield Agitation Inventory. See: https://www.cambridge.org/core/journals/intern ational-psychogeriatrics/article/abs/conceptualization-of-agitation-results-based-on-the-cohenm ansfield-agitation-inventory-and-the-agitation-behavior-mapping-instrument/36F895AFD524673 CA46B3F7294A78F50.

[14] See: Sahlen et al. [76].

This study involved 72 patients divided into two groups: the intervention group (n = 36) that received person-centred and integrate PREFER care over 6 months, and the control group (n = 36) that received standard health care recently provided by a primary healthcare centre or the led by a nurse's heart clinic at the hospital.

To assess health-related quality of life, the 5Q-ED instrument (five questions) was used to calculate quality-adjusted life years (QALYs). In this study, assessment only directs cost from the provider perspective. To avoid double-counting in cost assessment, patients' costs, indirect costs, and the expenses of state authorities are excluded. The main results showed that the intervention group had a slight improvement in QALYs (+0.006) compared to the control group with a slight decrease in QALYs (−0.024) (Ibid.). Also, the cost assessment results showed that over six months of intervention, costs were reduced to SEK600.000 (€61.000) according to the primary analysis, and according to sensitive analysis, costs were reduced by €49.000 (Ibid).

A recent study has also shown that home-based palliative care effectively reduces severe CHF patients' hospitalisation, but the cost-savings were not evaluated [94].

Results of the implementation strategies introduced in this section characterize major advancements in some aspects of health care. However, all three interventions need to invest significant efforts for progress in the effectiveness and cost-effectiveness of the interventions in the future. The examined health care outcomes of intervention also need to move from healthcare facilities and healthcare system perspectives to patients' perspectives if we want to have a person in the centre of healthcare systems.

6.5 Moving Forward: Valuing Intangibles in Healthcare[15]

Today, administrative goals have taken over clinical goals (mainly in the form of cost-cutting). The way clinical goals are achieved also changed: the patient is considered as a consumer (since the 1960s in the US, growing in France). Doctors are not as legitimate as they used to, because a paternalistic approach is no longer advocated for. They have to take into account the patient's perspective, values and requests. This change in the power balance, as well as the uncertainty still attached to care despite tremendous technological and medical progress, appeals for reinstalling relationships as a core intangible component of care.

The reflection on the value of intangibles starts with the prevalent belief in the business and public policy arenas that "it you can't measure it, you can't fix it". Therefore, as Pierron and Vinot put it: "*there is a need, at a time when standardized quantitative approaches have come to dominate the care system, of compensating for this through attention to more qualitative, even phenomenological, data: narration as*

---

[15] This section partly incorporates a paper presented at European Group for Organization Studies: Vinot and Chelle [93], "The evaluation of relational value in health care organizations: A conceptual framework", EGOS 2018.

opposed to classification; the personal as opposed to the personalized; the individual as opposed to the individuated" [69]. How can we choose and create appropriate indicators in this logic?

The criticisms addressed to Patient-reported Outcomes (requested by the provider or the industry to ensure compliance with regulations and manage reimbursement schemes), leading to Patient-reported Information (feedback sent by patients via the Internet), reveals the limits of traditional measurements techniques to incorporate narrative components into evaluation [77]. For instance, if a doctor sees part of his remuneration modulated according to the number of complaints filed against him, this does not automatically imply that he will improve the quality of care provided to avoid these claims. It can also deter patients from starting any procedure, for example by persuading them that they are useless. The authors advocate for a narrative version of PRI, which cannot be reduced to a series of metrics. Nevertheless, to overcome methodological difficulties, the authors endorse regulating payment models on what matters for patients (patient-valued outcomes). Rather than a culture of results, the study encourages a culture of learning for practitioners based on patient experience. The authors conclude, however, by stressing the idiosyncratic nature of the results thus produced. This implies looking closely at contexts of collective healthcare activities in which human and nonhuman interactions in a search of coordination of expertise and values are constitutive of outcomes [23].

Management science offers tools to integrate the unmeasurable into business strategies. For instance, quantifying intangible assets such as people, information and customer relationships was the principle of the Balanced Scorecard [44]. With that original management tool, the authors asserted that performance could not be measured only by economic results. That last statement certainly applies to health. However, when we seek to endorse a more phenomenological stance incorporated into healthcare activity, we need methods that reconcile both evidence-based medicine (measurable proofs and outcomes) and narrative medicine that the move to person-centered care have initiated: patients are also considered as agents and partners [70] and sensory-care is also crucial to privilege "conflicts of interpretations" among the accuracy of proof [11].

How could we value intangibles from a management science perspective? The notion of intangibles has gained considerable traction in the finance sector. It accounts for the mechanisms by which value is created on markets. The definition provided for the financial sector—an organizational and relational capacity based on skills and knowledge [90]—can be imported to the health sector. Managing intangible activities appears to be the main way to add value and a possible solution to evaluate the relational value in healthcare organizations. What cannot be counted matters not only from a clinical perspective, but also from a management perspective. Health professionals already know the benefits of the relational dimension of care, hence we have to reach out to the managers in their own language to fully implement a person-centered perspective.

Table 6.4 lays down the draft of a model aiming at capturing relational value. It draws from an organizational and empirical perspective, rather than an economic and formal one. Relational value can be observed not only between two or more people

Table 6.4 An observation guide for evaluating relational value

Observation level	Relational units	Evaluation tools
Interpersonal	Trusting collaboration Problem-solving mindset Empathic "communicaction" Acceptance of differences	Patient-reported information (narrative feedback)
Organizational	User-friendly time and space Circulation of reliable information Safe and effective clinical processes Conflict equilibrium	Researcher's log based on observational study
Environmental	Partnerships Reputation Frequentation Adequate transportation system	Mapping

in the same room (interpersonal level), but works at the organization level and environment level. Promoting relational value thus does not imply an individualistic or atomistic perspective on care. We propose this model for evaluating relational value. The corresponding evaluation tools draw from an enriched qualitative methodology, based on narrative enquiry [50, 57], observational study [83] and mapping [44].

At the interpersonal level, the global technical competence of professionals has to be complemented by various sorts of relational skills. A trusting collaboration may be the first criteria. That entails reinforcing team-work that has been weakened in hospitals settings in the name of interchangeability of agents. It is detrimental to care in the sense that it deters mutual adjustments between healthcare professionals. When team members know each other, collaboration adds more value than just cooperation. In healthcare organizations, professionals are used to applying protocols. Each category of healthcare professionals follows rules and regulations in their specialty. That can result in disjunctions in everyday practice, because situations can require transversal actions, hence the need for a problem-solving mindset in the healthcare workplace. Of course, cognitive agility does not set aside routines altogether. Nonetheless adaptation should prevail over planning and protocols, because it can make a difference for the patient. Communication is a well-known component of a good relationship with the patient. Empathic "communic-action" would mean not only listening and talking, but also taking action when necessary (concerning pain relief, for instance). Managers should not overlook that dimension as wasted time, but could appraise it as invested time for more efficacy in the healing and/or caring journey. It draws from a reflex of asking oneself what can help that person to carry on, what Mintzberg called "judgment". Each patient is different, so taking into consideration the person's needs is essential, Economists speak about "preferences." Sociologists and philosophers prefer the word "values." Although those terms are not exactly interchangeable, what we must focus on is the word "person" instead of "patient." Sickness is not the only lens through which we should see the human being sitting in a healthcare organization. For that, it is crucial to put oneself in a position of acceptance of alterity, and

"acceptance" simultaneously meaning "acknowledgment" and "belief in the good-
ness of something." Otherness can trigger a rejection in the absence of an appropriate
training. Overcoming that feeling can be challenging; therefore, efforts have to be
undertaken from a long-term perspective. Overall, one can say that interpersonal
interaction should be guided by the following principle: "to achieve real quality
in health care, we require personalized services on a human scale, not impersonal
interventions on an economic scale." [58].

If relations take place at the interpersonal level, that does not mean that they
happen in the vacuum. Improvements can be worked on at the organizational level
to induce a greater relational value. The spatial organization of healthcare has also
been insufficiently undermined.

User-friendly space and time enhances relationships in and around care. Architecture can induce positive and fruitful encounters (with things as mundane as the
location of a coffee machine, the position of the bed in the patient's room, or the
colors on the walls). User-friendly space sustains user-friendly time, meant as quality
time. Time is a scarce resource in healthcare organization. It has been made scarce
by management techniques commending more productivity. Cost-efficiency analysis
uses quantitative indicators of outcomes [38], such as the amount of time a patient
stays in a service or occupies a room. If health professionals cannot take the time,
are they not compelled to miss the point of care? Quality time is not wasted time,
as it can help the patient to feel better, and the professional to coordinate better.
Organizational schemes should also include quality time with health professionals
and families. Waiting rooms and meeting points cannot be limited to an assigned
space. Quality time derives from quality space. The circulation of reliable information happens through formal and informal circuits. Information systems should
make patients' data available to different services within the organization. Within
hospitals, the dual hierarchy, clinical and administrative [15], create discrepancies
in processes. Clinical goals remain the priority in a healthcare settings. The organization is reliable as long as it ensures safe and effective clinical processes, which
calls for clear routines as well as adaptive strategies. "Effective" is preferable to
"efficient," the latter belonging to the vocabulary of a machine organization, where
healthcare organizations are professional bureaucracies. The term "effective" is the
one to promote intangible values: "what people call efficiency all too often reduced to
economy, more specifically to economizing: cutting tangible costs at the expense of
intangible benefits" [58]. On the management front, a conflict equilibrium has to be
contained. In large organizations, conflicts have to be handled, but cannot be avoided.
Some conflicts can paradoxically motivate teams to work together or leaders to take
action. Constant unresolved tensions, on the contrary, create a climate of hostility
and induce exit behaviours.

Last but not least, intangible values need to be taken at the environmental level.
Partnerships are an extent of cooperation at the community level. These can be either
institutional (cooperation between health organizations, community-based organizations, primary care doctors and units…) or virtual (online patient communities).
Reputation is built within those networks. Next to the official rankings, word of mouth
is essentially relational and matters to institutions. When people get to choose their

place of care, a hospital or clinic's reputation gives an edge. To guarantee a satisfying relational quality, there is an optimal number of beds to consider, enough to cover baseline costs, not too many so as not oversize buildings and keep distances short. Frequentation has to be optimal, not maximized. Finally, an adequate transportation system should not be overlooked. Easy connections to hospitals and care facilities are an important factor for the patient being visited by their family and friends.

6.6   Conclusion

The objective of this chapter was to provide conceptual understandings of evaluation methods for healthcare and concrete illustrations so as to take stock on advancements and applications. As a conclusive reflection, our understanding of an effective and ethical patient-centered healthcare system consists in considering the person as a whole and creating the conditions to make them a visible and proactive subject in the care journey. Going from macro to micro levels of evaluation of healthcare in this chapter, we showed that a lot more can be observed or deducted from the material organization of care, which enables consideration of the intangibles, notably *what has value* is not necessarily *what is worth* doing to be considerate of the patient in the care process. Would going forward mean going back to an ancient wisdom? The well-known doctor Francis Peabody [67] wrote almost a century ago:

> The good physician knows his patients through and through, and his knowledge is bought dearly. Time, sympathy and understanding must be lavishly dispensed, but the reward is to be found in that personal bond which forms the greatest satisfaction of the practice of medicine. One of the essential qualities of the clinician is interest in humanity, for the secret of the care of the patient is in caring for the patient. [67]

Then, at that time evaluation existed, but it was not the massive trend as we know it today. The words "judgment", "appreciation" and "worth" were still prevalent over "calculus", "evaluation" and "value". Our bottom line is not to go back in time. It is to reinstall observation as a valid tool to support the patient in and around care. A century ago, the principles expressed by Dr. Peabody were conceived for acute care. Today, with chronic conditions and longer lives, we should strengthen, not set aside, the value of relationships in care.

Acknowledgements   This publication is based upon work from COST Action "**European Network for cost containment and improved quality of health care-CostCares**" (CA15222), supported by COST (European Cooperation in Science and Technology)

COST (European Cooperation in Science and Technology) is a funding agency for research and innovation networks. Our Actions help connect research initiatives across Europe and enable scientists to grow their ideas by sharing them with their peers. This boosts their research, career and innovation.

https://www.cost.eu

References

1. Adam, A., Delis, M., Kammas, P.: Public sector efficiency: leveling the playing field between OECD countries. Public Choice **146**(1/2), 163–183 (2011)
2. Afonso, A., St. Aubyn, M.: Relative efficiency of health provision: a DEA approach with non-discretionary inputs, ISEG-UTL Economics Working Paper No. 33/2006/DE/UECE (2006) (https://papers.ssrn.com/sol3/papers.cfm?abstract_id=952629)
3. Anton, S. G., Onofrei, M.: Health care performance and health financing systems in countries from Central and Eastern Europe. Transylvanian Rev. Administr. Sci. **35** E, 22–32 (2012)
4. Baldwin, M., Spong, A., Doward, L., Gnanasakthy, A.: Patient-reported outcomes, patient-reported information. Patient: Patient-Centered Outcomes Res. **4**, 11–17 (2011)
5. Banke-Thomas, A., Nieuwenhuis, S., Ologun, A., Mortimore, G., Mpakateni, M.: Embedding value-for-money in practice: a case study of a health pooled fund programme implemented in conflict-affected South Sudan. Eval. Program Plann. **77**, 101725 (2019)
6. Berger, M.C., Messer, J.: Public financing of health expenditures, insurance, and health outcomes. Appl. Econ. **34**, 2105–2113 (2002)
7. Bhat, V.N.: Institutional arrangements and efficiency of health care delivery systems. Europ. J. Health Econ. **50**, 215–222 (2005)
8. Block, D. J.: Healthcare outcomes management: Strategies for planning and evaluation. Jones and Bartlett, Sudbury, MA (2006)
9. Brännström, M., Boman, K.: Effects of person-centred and integrated chronic heart failure and palliative home care. PREFER: A Randomized Controlled Study. Eur. J. Heart Fail. **16**(10), 1142–1151 (2014)
10. Bui, A.L., Horwich, T.B., Fonarow, G.C.: Epidemiology and risk profile of heart failure. Nat. Rev. Cardiol. **8**(1), 30–41 (2011)
11. Chvetzoff, R., Chvetzoff, G., Pierron, J.P.: Using tools for plotting the right course: evidence Based Medicine or relationship-centered care? Reflections based on the recommendation for autism published by the French National Health Authority (Haute Autorité de santé). Ethique & Santé **9**(4), 159–164 (2012)
12. de Cos, P.H., Moral-Benito, E.: Determinants of health-system efficiency: evidence from OECD countries. Int. J. Health Care Finance Econ. **14**, 69–93 (2014)
13. Chen, A., Sabharwal, S., Akhtar, K., Makaram, N., Gupte, C. M.: Time-driven activity based costing of total knee replacement surgery at a London teaching hospital. The Knee **22**, 640–645 (2015)
14. Cheng, Y., Raisch, D. W., Borrego, M. E., Gupchup, G. V.: Economic, clinical, and humanistic outcomes (ECHOs) of pharmaceutical care services for minority patients: a literature review. Res. Soc. Admin. Pharmacy **9**, 311–329 (2013)
15. Cockerham, W.C.: Max Weber: Bureaucracy, formal rationality and the modern hospital. In: Collyer, F. (ed.) The Palgrave Handbook of Social Theory in Health, Illness and Medicine, pp. 124–140. Palgrave Macmillan, New York, N.Y. (2015)
16. Crott, R., Lawson, G., Nollevaux, M.-C., Castiaux, A., Krug, B.: Comprehensive cost analysis of sentinel node biopsy in solid head and neck tumors using a time-driven activity-based costing approach. Eur. Arch. Otorhinolaryngol. **273**, 2621–2628 (2016)
17. Drucker, P. F.: The New Realities: In Government and Politics, in Economics and Business, in Society and World View. Harper & Row, New York (1989)
18. Drummond, M.F.: Methods for the economic evaluation of health care programmes. Oxford University Press, Oxford (1987)
19. Drummond, M.F., Sculpher, M.J., Claxton, K., Stoddart, G.L., Torrance, G.W.: Methods for the economic evaluation of health care programmes. Oxford University Press (2015)
20. Eit Health.: Implementing Value-Based Health Care in Europe: Handbook for Pioneers (2020)
21. Ekman, I., Wolf, A., Olsson, L.-E., Taft, C., Dudas, K., Schaufelberger, M., Swedberg, K.: Effects of Person-centred care in patients with chronic heart failure: the PCC-HF study. Eur. J. Heart Fail. **33**(9) (2012)

22. El Alaoui, S., Lindefors, N.: Combining time-driven activity-based costing with clinical outcome in cost-effectiveness analysis to measure value in treatment of depression. PLOS ONE **11**, e0165389 (2016)
23. Engeström, Y.: Expertise in Transition: Expansive Learning in Medical Work. Cambridge University Press (2018)
24. European Commission.: Defining value in "value-based healthcare" - Report of the Expert Panel on effective ways of investing in Health (EXPH). Luxembourg: European Union (2019)
25. Evans D. E., Tandon A., Murray C. J. L., Lauer J. A.: Comparative efficiency of national health systems: cross national econometric analysis. BMJ: British Med. J. **323**, 307–310 (2001)
26. Fleming, F.: Evaluation methods for assessing Value for Money. Australasian Evaluation Society (2013). http://betterevaluation.org/sites/default/files/Evaluating%20methods%20for%20assessing%20VfM
27. Frogner, B. K., Frech H.E., Parente, S.T.: Comparing efficiency of health systems across industrialized countries: a panel analysis, BMC Health Services Res. **15**, 415 (2015)
28. Giuffrida, A., Gravelle, H.: Measuring performance in primary care: econometric analysis and DEA. Appl. Econ. **33**, 163–175 (2001)
29. Glouberman, S., Mintzberg, H.: Managing the care of health and the cure of disease - Part I: Differentiation. Health Care Manag. Rev. **26**, 56–69 (2001)
30. Glouberman, S., Mintzberg, H.: Managing the care of health and the cure of disease - Part II: Integration. Health Care Manag. Rev. **26**, 70–84 (2001)
31. Gold, M.R., Siegel, J.E., Russell, L.B., Weinstein, M.C., Russell, L.B.: Cost-Effectiveness in Health and Medicine. Oxford University Press (1996)
32. Golden, B.: Transforming Healthcare Organizations. Healthc. Q. **10**, 10–19 (2006)
33. Goncharuk, A.G.: Socioeconomic criteria of healthcare efficiency: an international comparison. J. Appl. Manag. Invest. **6**, 89–95 (2017)
34. Goncharuk, A.G.: Wine business performance benchmarking: a comparison of German and Ukrainian wineries. Benchmarking: An Int. J. **25**, 1864–1882 (2018)
35. González, E., Cárcaba, A., Ventura, J.: Value efficiency analysis of health systems: does public financing play a role? J. Public Health **18**, 337–350 (2010)
36. Gregório, J., Russo, G., Lapão, L.V.: Pharmaceutical services cost analysis using time-driven activity-based costing: a contribution to improve community pharmacies' management. Res. Social Adm. Pharm. **12**, 475–485 (2016)
37. Groenewoud, A.S., Westert, G.P., Kremer, J.A.M.: Value based competition in health care's ethical drawbacks and the need for a values-driven approach. BMC Health Serv. Res. **19**, 256 (2019)
38. Gusmano, M.K., Kaebnick, G.: Clarifying the role of values in cost-effectiveness. Health Econ. Policy Law **11**(4), 439–443 (2016)
39. Hadad, S., Hadad, Y., Simon-Tuval, T.: Determinants of healthcare system's efficiency in OECD countries. Eur. J. Health Econ. **14**, 253–265 (2013)
40. Hollingsworth, B.: Non-parametric and parametric applications measuring efficiency in health care. Health Care Manag. Sci. **6**, 203–218 (2003)
41. Hollingsworth, B., Wildman, J.: The efficiency of health production: re-estimating the WHO panel data using parametric and non-parametric approaches to provide additional information. Health Econ. **12**, 493–504 (2002)
42. ICAI.: ICAI's Approach to Effectiveness and Value for Money. Independent Commission for Aid Impact (ICAI) (2011)
43. ICAI.: DFID's approach to value for money in programme and portfolio management: A performance review (2018)
44. Kaplan, R.S., Norton, D.P.: Strategy Maps: Converting Intangible Assets into Tangible Outcomes. Harvard Business School Press, Boston, MA (2010)
45. Kaplan, R.S., Porter, M.E.: How to solve the cost crisis in health care. Harv. Bus. Rev. **89**(9), 46–52 (2011)

46. Kaplan, R.S., Witkowski, M., Abbott, M., Guzman, A.B., Higgins, L.D., Meara, J.G., Padden, E., Shah, A.S., Waters, P., Weidemeier, M., Wertheimer, S., Feeley, T.W.: Using time-driven activity-based costing to identify value improvement opportunities in healthcare. J. Healthc. Manag. **59**, 399–412 (2014)

47. Keel, G., Savage, C., Rafiq, M., Mazzocato, P.: Time-driven activity-based costing in health care: a systematic review of the literature. Health Policy **121**, 755–763 (2017)

48. Kim, Y., Kang, M.: The measurement of health care system efficiency: cross-country comparison by geographical region. The Korean J. Policy Stud. **29**, 21–44 (2014)

49. Kim, Y., Oh, D., Kang, M.: Productivity changes in OECD healthcare systems: bias-corrected Malmquist productivity approach. Int. J. Health Plann. Manag. **31**, 537–553 (2016)

50. Kleinman, A.: The Illness Narratives: Suffering, Healing, and the Human Condition. Basic Books, New York (1988)

51. Lavallee, D.C., Chenok, K.E., Love, R.M., Petersen, C., Holve, E., Segal, C.D., Franklin, P.D.: Incorporating patient-reported outcomes into health care to engage patients and enhance care. Health Aff. **35**, 575–582 (2016)

52. Lorenzoni, L., Murtin, F., Springare, L.-S., Auraaen, A. Daniel, F.: Which Policies Increase Value for Money in Health Care? OECD Health Working Papers. No. 104, OECD Publishing, Paris (2018)

53. Lord, L., Gale, N.: Subjective experience or objective process: understanding the gap between values and practice for involving patients in designing patient-centred care. J. Health Organ. Manag. **28**(6), 714–730 (2014)

54. Lo Storto, C., Goncharuk, A.G.: Efficiency vs effectiveness: a benchmarking study on European healthcare systems. Econ. Sociol. **10**, 102–115 (2017)

55. Mackinnon Iii, G. E.: Understanding Health Outcomes and Pharmacoeconomics. Jones & Bartlett Publishers (2011)

56. Martini, G., Berta, P., Mullahy, J., Vittadini, G.: The effectiveness–efficiency trade-off in health care: the case of hospitals in Lombardy, Italy. Reg. Sci. Urban Econ. **49**, 217–231 (2014)

57. Marini, M.G.: Narrative Medicine: Bridging the Gap Between Evidence-Based Care and Medical Humanities. Springer, Cham (2016)

58. Mintzberg, H.: Managing the Myths of Healthcare: Bridging The Separations Between Care, Cure, Control and Community. Berrett-Koehler Publishers, Oakland, CA (2017)

59. Mjåset, C., Ikram, U., Nagra, N. S., Feeley, T. W.: Value-Based Health Care in Four Different Health Care Systems. NEJM Catalyst Innovations in Care Delivery, 1 (2020)

60. National Institute for Health and Care Excellence.: Multimorbidity: Clinical Assessment and Management. (2016) (https://www.nice.org.uk/guidance/ng56/resources/multimorbidity-clinical-assessment-and-management-pdf-1837516654789).

61. National Institute for Health and Care Excellence.: Guide to the Methods of Technology Guide to the Methods of Technology appraisal 2013 (2013) https://www.nice.org.uk/process/pmg9/resources/guide-to-the-methods-of-technology-appraisal-2013-pdf-2007975843781

62. Nilsson, K., Bååthe, F., Andersson, A.E., Wikström, E., Sandoff, M.: Experiences from implementing value-based healthcare at a Swedish University Hospital—A longitudinal interview study. BMC Health Serv. Res. **17**, 169–169 (2017)

63. Nolte E., Wait S., McKee, M.: Investing in health: benchmarking health systems, technical report published by The Nuffield Trust, London (2006)

64. OECD.: Health care systems: getting more value for money. Author Paris, France (2010)

65. OECD: Geographic variations in health care: what do we know and what can be done to improve health system performance? OECD Health Policy Stud. (2014). https://doi.org/10.1787/9789264216594-en

66. Papanicolas, I., Smith, P.C. (eds.): Health systems performance comparison: an agenda for policy, information and research. Open University Press, USA, New York (2013)

67. Peabody, F.: The care of the patient. J. Am. Med. Assoc. **88**, 877–882 (1927)

68. Pendleton, R.: We won't get value-based health care until we agree on what "value" means. Harv. Bus. Rev. **2**, 2–5 (2018)

69. Pierron, J.P., Vinot, D.: The meaning of value in "Person-centred" approaches to Healthcare. Eur. J. Pers. Cent. Healthc. **8**(2), 193–200 (2018)
70. Pirhonen, L., Gyllensten, H., Olofsson, E.H., Fors, A., Ekman, I., Bolin, K.: The cost-effectiveness of person-centred care provided to patients with chronic heart failure and/or chronic obstructive pulmonary disease. Health Policy OPEN **1**, 1–7 (2020)
71. Porter, M.E., Olmsted Teisberg, E.: Redefining health care: Creating value-based competition on results. Harvard Business School Press, Boston, MA (2006)
72. Porter, M.E.: What is value in health care? N. Engl. J. Med. **363**(26), 2477–2481 (2010)
73. Porter, M.E., Kaplan, R.S.: How to pay for health care. Harv Bus Rev. **94**, 88–100 (2016)
74. Porter, M. E., Lee, T. H.: The strategy that will fix health care. Harvard Business Review (2013)
75. Quinn, R.E., Rohrbaugh, J.: A spatial model of effectiveness criteria: towards a competing values approach to organizational analysis. Manag. Sci. **29**(3), 363–377 (1983)
76. Sahlen, K.-G., Boman, K., Brännström, M.: A cost-effectiveness study of person-centered integrated heart failure and palliative home care: based on a randomized controlled trial. Palliat. Med. **30**(3), 296–302 (2016)
77. Schlesinger, M., Grob, R., Shaller, D.: Using patient-reported information to improve clinical practice. Health Serv. Res. **50**(S2), 2116–2154 (2015)
78. Siguenza-Guzman, L., Auquilla, A., Van Den Abbeele, A., Cattrysse, D.: Using time-driven activity-based costing to identify best practices in academic libraries. J. Acad. Librariansh. **42**, 232–246 (2016)
79. Smith, P.C.: Measuring Value for Money in Healthcare: Concepts and Tools. The Health Foundation, London (2009)
80. Smith, P.C.: What is the scope for health system efficiency gains and how can they be achieved? Eurohealth **18**(3), 3–6 (2012)
81. Soderland, N., Kent, J., Lawyer, P. & Larsson, S. (2012). Progress towards value-based health care. Lessons from 12 Countries. The Boston Consulting Group. Inc.
82. Sorenson, C., Drummond, M., Kanavos, P.: Ensuring value for money in health care: the role of health technology assessment in the European Union. WHO Regional Office Europe (2008)
83. Strauss, A.L.: Negotiations: Varieties, Contexts, Processes, and Social Order. Jossey-Bass, San Francisco (1978)
84. Surr, C.A., Holloway, I., Walwyn, R.E.A., Griffiths, A.W., Meads, D., Martin, A., Kelley, R., Ballard, C., Fossey, J., Burnley, N., Chenoweth, L., Creese, B., Downs, M., Garrod, L., Graham, E.H., Lilley-Kelly, A., McDermid, J., McLellan, V., Millard, H., Perfect, D., Robinson, L., Robinson, O., Shoesmith, E.N.S., Stokes, G., Wallace, J., Farrin, D.A.: Effectiveness of Dementia Care Mapping™ to reduce agitation in care home residents with dementia: an open-cohort cluster randomised controlled trial. Aging Ment. Health **13**(1), 1–14 (2020)
85. Tak, H., Ruhnke, G.W., Shih, Y.T.: The association between patient-centered attributes of care and patient satisfaction. Patient: Patient-Centered Outcomes Res. **8**, 187–197 (2015)
86. Tchouaket, É.N., Lamarche, P.A., Goulet, L., Contandriopoulos, A.P.: Health care system performance of 27 OECD countries. Int. J. Health Plann Manag. **27**, 104–129 (2012)
87. Teisberg, E., Wallace, S., O'hara, S.: Defining and implementing value-based health care: a strategic framework. Acad. Med. J Assoc. Am. Med Colleges **95**, 682–685 (2020)
88. The Economist Intelligence Unit.: Value-based healthcare: A global assessment Findings and methodology (2016)
89. Thorn, J., Man, M.-S., Chaplin, K., Bower, P., Brookes, S., Gaunt, D., Fitzpatrick, B., Gardner, C., Guthrie, B., Hollinghurst, S., Lee, V., Mercer, S.W., Salisbury, C.: Cost-effectiveness of a patient-centred approach to managing multimorbidity in primary care: a pragmatic cluster randomised controlled trial. BMJ Open **10**(1), 1–10 (2019)
90. Vallejo Alonso, B., Rodríguez Castellanos, A., Arregui-Ayastuy, G.: Identifying, measuring, and valuing knowledge-based intangible assets: new perspectives. IGI Global, Hershey, PA (2011)
91. Varabyova, Y., Schreyögg, J.: International comparisons of the technical efficiency of the hospital sector: Panel data analysis of OECD countries using parametric and non-parametric approaches. Health Policy **112**, 70–79 (2013)

92. Verma, R.: Overview: what are PROMs and PREMs. NSW Agency for Clinical Innovation (2016)
93. Vinot D., Chelle, E.: The evaluation of relational value in health care organizations: a conceptual framework. Conference paper at EGOS 2018 (2018)
94. Wong, R.C., Tan, P.T., Aziz, S., Seow, Y.H.: Home-based advance care programme is effective in reducing hospitalisations of advanced heart failure patients: a clinical and healthcare cost study. Ann. Acad. Med. **42**(9), 466–471 (2013). PMID: 24162321
95. World Bank.: Database: Health Indicators (2020). http://data.worldbank.org/indicator/Data. Accessed 20 December 2020
96. World Health Organization.: Health systems: Improving performance. The World Health Report (2000)
97. World Health Organization: Monitoring the Building Blocks of Health Systems: A Handbook of Indicators and their Measurement Strategies. WHO Press, Geneva (2010)
98. Yu, Y.R., Abbas, P.I., Smith, C.M., Carberry, K.E., Ren, H., Patel, B., Nuchtern, J.G., Lopez, M.E.: Time-driven activity-based costing to identify opportunities for cost reduction in pediatric appendectomy. J. Pediatr. Surg. **51**, 1962–1966 (2016)

Chapter 7
An Overview of Measurement Systems and Practices in Healthcare Systems Applied to Person-Centred Care Interventions

Vítor Raposo, Darijana Antonić, António Casa Nova, Roman Andrzej Lewandowski, and Paulo Melo

Abstract Person-centred care (PCC) is an increasing international priority and a shift in health systems orientation and development. Innovative models are required across Europe to prototype healthcare based on health promotion and PCC to improve healthcare quality and costs containment. Regardless of the type of intervention, investments will be required, and it will be essential to demonstrate the value created, comparing consequences and the associated costs. Independent of PCC intervention, we must consider different decision levels and stakeholders in the process. This work aims to focus on a broader perspective of health governance on PCC implementations, considering the need and importance of measurement systems (outcomes and costs) to support and evaluate innovative health service delivery models. It is necessary to have a global view of the entire system considering, from a health governance perspective, the different decision-making levels, the multiple stakeholders and

V. Raposo (✉)
Faculty of Economics, Centre for Business and Economics Research (CeBER), Centre of Health Studies and Research, University of Coimbra, Av. Dr. Dias da Silva 165, 3004-512 Coimbra, Portugal
e-mail: vraposo@fe.uc.pt

D. Antonić
Public Health Institute of the Republic of Srpske, Jovana Dučića 1, 78000 Banja Luka, Republic of Srpska, Bosnia and Herzegovina

A. C. Nova
Health School of Polytechnic Institute of Portalegre, Campus Politécnico 10, 7300-555 Portalegre, Portugal
e-mail: casanova@ipportalegre.pt

R. A. Lewandowski
Institute of Management and Quality Science, Faculty of Economics, University of Warmia and Mazury, 10-720 Olsztyn, Poland
e-mail: rlewando@wp.pl

P. Melo
Faculty of Economics, Centre for Business and Economics Research, INESC Coimbra, University of Coimbra, Av. Dr. Dias da Silva 165, 3004-512 Coimbra, Portugal
e-mail: pmelo@fe.uc.pt

© The Author(s) 2022
D. Kriksciuniene and V. Sakalauskas (eds.), *Intelligent Systems for Sustainable Person-Centered Healthcare*, Intelligent Systems Reference Library 205,
https://doi.org/10.1007/978-3-030-79353-1_7

the alignment of their interests. Value-Based Healthcare (VBHC), Value for Money (VfM) and economic evaluation provide concepts, methodologies, and tools that can be used to compare costs and consequences evaluating their impact on society. We need accurate outcomes and costs measurement systems and evaluation tools that can be incorporated in an organizational environment supporting organizational learning and interaction in exchanging knowledge and experience about implementation.

Keywords Health governance · Measurement systems · Person-centred care frameworks · Value for money · Economic evaluation · Value-based healthcare · Costs · Outcomes

7.1   Introduction

Ensuring that healthcare is person-centred is an increasing international priority and a shift in orientation in health systems and its development. The World Health Organisation (WHO) has promoted and supported this approach in its global strategy for people-centred and integrated health services, and the Organisation for Economic Co-operation and Development (OECD), also confirmed their strong support in these efforts [58]. This aim implies that services are organized around people needs and expectations to make them more socially relevant and responsive while producing better outcomes [87, 88]. Besides the active involvement of the different stakeholders, it also empowers people to have a more active role in their health with broader benefits to individuals and their families, health professionals, communities, and health systems [89].

There is evidence that Person-centred care (PCC) improves health outcomes, maintains health care quality, and helps cost containment [46]. It is also a breeding field for innovation [21, 37, 38, 60] that can improve coordination and access to health care and services [10, 75] and other effects to the whole system. According to Louw et al. [48], PCC practice brings with it a broad number of benefits for the patient (higher satisfaction, improved health and quality of care, additional preventive care, better functional performance, and increased engagement), for the healthcare system (better adherence to treatment, recommendations and follow-up visits, increased Efficiency of care, reduction of the number of hospitalizations and shorter hospital stays), and the clinician (more satisfaction, better time management, and fewer complaints from patients).

The development of specific frameworks to capture the PCC essence and healthcare processes and outcomes has been extended to both generic and field-specific care delivery approaches. In a critical review of literature related to PPC, Phelan et al. [63] identified several frameworks, some in specific fields—like dementia, older person services, and acute care settings [22, 30]—and other developed as generic frameworks—like the Santana et al. [72] framework based on Donabedian model for healthcare improvement, and the Person-centred Practice Framework [50–52] with the broadest applicability and which has been most extensively adopted. The frameworks cited in the review present examples used in different countries and that could be used as a roadmap for PCC implementation. However, all of these frameworks have strengths and limitations, and offer variable utility for healthcare practice.

Innovative models are required across Europe to prototype healthcare based on health promotion and PCC, including Health Labs development to improve healthcare quality and costs containment [46]. Nevertheless, in the context of managing constrained healthcare budgets while safeguarding equity, access and choice, governments face the challenge to strategically manage scarce resources by investing in interventions that provide the best health outcomes [7, 47, 57, 80]. Regardless of the type of intervention, investments will be required, and as such, it will be essential to demonstrate the value created, comparing consequences and the associated costs. Independent of the type of PCC intervention, sponsors, policymakers, and managers require evaluations that prove that the additional health care resources needed to make the procedure, service, or program available to those who could benefit from it are justified.

Value-Based HealthCare [64, 66], Value for Money [29, 78] and economic evaluation [19, 31] all provide concepts, methodologies and tools to look at costs and consequences. They also, reinforce accountability assuring that taxpayers' and sponsors' money is spent wisely and that the stakeholder's interests are safeguarded [29, 41, 78, 80]. To use the different instruments provided from these approaches in PCC settings, a robust measurement approach is needed. Silva [77] mentioned that such an approach helps to understand the extent to which care is person-centred and helps to differentiate worthwhile initiatives. The investment in measures that help to assess whether health systems deliver what matters to people, and not only how much they cost, is being recognized as a meaningful action to reorient health systems to become more people-centred [58]. This approach is also extensible for the costs side. According to Kaplan et al. [43], the currently existing cost systems in healthcare rely on inaccurate and arbitrary cost allocations and provide little transparency to guide first-line care providers attempts to understand and change the proper drivers of their costs.

Based on the issues mentioned, this chapter aims to focus on a broader perspective of health governance on PCC implementations, considering the need and importance of measurement systems (outcomes and costs) to support and evaluate new innovative models of health service delivery.

We start by introducing the concept of health governance and its different decision levels, using the Santana et al. [72] framework based on Donabedian model for healthcare improvement, and the VfM framework proposed by ICAI [41]. Taking these frameworks together, in our opinion, allows a global perspective of implementation, costs and consequences from a health governance point of view that can be applied at different levels of decision-making.

Next, we will look to the components of value according to Porter [64], measuring health outcomes that matter to patients and the cost of delivering the outcomes. Using several references, we will consider the importance of metrics on PCC, different types of person outcomes, and sources of measures, instruments and indicators that can be used to this end. Finally, we focus on the involved costs, namely different types and the difficulties in obtaining them, and possible solutions to get them in a more accurate way. We finish with a small real-life example of the application of a measurement framework, highlighting advantages and drawbacks.

7.2   Person-Centred Care Governance, Frameworks for Implementation and Value?

Health governance is defined as the actions and means adopted by society to organize itself to promote and protect its population's health [17]. This approach devotes special attention to strengthening the capacity to formulate policies, develop good governance systems, set priorities at various levels, strengthen and broaden partnerships for health, and implement evaluation and monitoring [86]. Health governance encompasses different decision-making levels with distinct characteristics, each with its separate group of decision-makers, and interacting with each other in complex patterns. There are usually national government decisions determining the basic structure, organization and finance models of the entire health care system and the health organizations within it. Considering the different characteristics of health systems and countries' various administrative autonomy levels for some policy decisions, the regional or local governments could also be involved. There is also decision-making at the overall institutional level of healthcare organizations. At a more micro level, it focuses on the day-to-day operational management of staff and services inside the health organizations.

The WHO [89] global strategy on integrated people-centred health services proposes five interdependent strategic directions: empowering and engaging people, strengthening governance and accountability, reorienting the model of care to more efficient and effective health care services, coordinating services around the needs of people at every level of care, enabling environment that brings together the different stakeholders to undertake the transformational change needed. This strategy means that a governance approach should be adopted to implement this global strategy at different levels. Although PCC must be implemented at the micro-level by care providers, PCC is also implemented at the meso (e.g., organizational) and macro (e.g., policy, organization and finance) service delivery levels of health systems.

One interesting framework that considers these different levels of implementation is proposed by Santana et al. [72]. The authors present a conceptual framework to guide systems and organizations in PCC provision and evaluation and health care quality improvement in general. This framework is based on the Donabedian [18] model domains of structure, process and outcome. It also encompasses different action levels (healthcare systems—organizations; patients—healthcare providers; patient—healthcare providers—healthcare systems) and identifies several constructs. Those elements are represented in Fig. 7.1 as domains, levels and constructs.

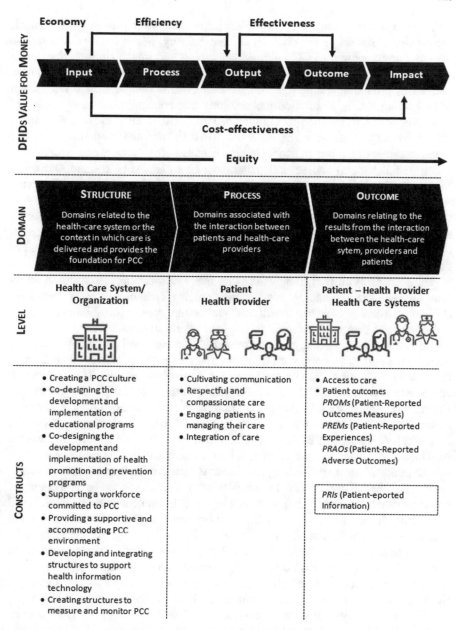

Fig. 7.1 Joining value for money and PCC—an integrated perspective. [71] Adapted from Santana et al. and ICAI [41]

There is a strong emphasis on the structure because it provides the context, the decisions that shape the context, of care delivery. Seven constructs compose the structure: creating a PCC culture, developing and implementing educational and training programs to health workers, given support to workforce committed with PCC, providing a supportive environment, the development and integration of supporting health information technologies, and creating structures to measure and monitor and PCC. The process includes constructs related to communication, human and respectful care, engaging patients in their care, and integrating care. The outcome dimension focus on access to care (timely access to care, care availability, and financial burden) and several patient outcomes (Patient-Reported Outcomes Measures—PROMs, Patient-Reported Experiences—PREMs, and Patient-Reported Adverse Outcomes—PRAOs).

The outcomes dimension proposed by Santana et al. [72] on the patient outcomes does not include a more recent source identified as data is patient-reported information (PRI) proposed by Baldwin et al. [6] that reinforce the patient perspective. For this reason, we included this category in the framework presented in Fig. 7.1.[1]

The Santana et al. [72] framework is an interesting roadmap, starting in structural pre-requisite and ending in outcomes, guiding systems and organizations in PCC solutions, evaluating and health care quality improvement combining evidence, guidelines, and best practice from existing frameworks implementation case studies. However, despite mentioning costs from different perspectives (costs related to access, cost reduction, costs with transport and medication, etc.) it does not guide how to deal with this component and relate it to the consequences. This framework can be extended if we introduce value-based healthcare (VBHC), value for money (VfM), and economic evaluation throughout the cycle since they provide concepts, methodologies, and tools to look at costs and consequences evaluating their impact.

Porter [64] presented VBHC as a healthcare delivery model in which providers are paid based on patient health outcomes. The value is calculated by measuring health outcomes that matter to patients against the cost of delivering the outcomes [64, 66], considering the full cycle of care for a particular medical condition. Several countries have embraced this agenda [54], with varying types of implementation [79, 82], and their principles guide the quest for high-quality health care for acceptable costs in health systems [36]. Although no country has fully implemented the VBHC agenda, it seems apparent that different theoretical framework elements function better in some healthcare systems than others [54]. Some critics nevertheless present doubts about the VBHC guide claims about being the appropriate way forward, supported by the existence of multiple understandings of what value for patients means [55, 61] and little availability of evidence related to VBHC effectiveness [45]. Groenewoud et al. [36] reinforce these points of view and argue that a single-minded focus on VBHC may cause serious infringements on medical ethical principles.

Recently, the European Institute of Innovation & Technology (EIT) Health published a handbook on how to adopt VBHC initiatives (EIT [20] that propose an implementation model entitled the VBHC Implementation Matrix, which defines five

---

[1] The concepts of PROMs, PREMs, PRAOs and PRIs will be expanded in the next section.

key dimensions critical to most VBHC initiatives: recording (measuring processes and outcomes through a scorecard and data platform), comparing (benchmarking teams through internal and external reports), rewarding (investing resources and creating outcome-based incentives), improving (organizing improvement cycles through collective learning),and partnering (aligning internal forces and forging collaborations with external partners). Focusing on one specific condition is an essential first step to maximising the implementation success, followed by a clear identification of resources needed and related outcomes. There is a need to look at the entire cycle, starting on the primary inputs and finishing on the resulting impact.

VfM is about obtaining the maximum benefit with the resources available. One proposed framework to VfM, proposed by the Department for International Development (DFID's) [41], is also included in Fig. 7.1. Their components are the input (staff, raw materials, capital), the process (the methods by which inputs are used), the outputs (what was delivery directly by agents), the outcome (what was achieved as a result), and the impact (long-term transformative change). The framework also compounds the four E's and cost-effectiveness [29, 40]: Economy is about minimizing the cost of inputs while considering quality, Efficiency focuses on achieving the best rate of conversion of inputs into outputs, again taking into mind quality, Effectiveness relates to achieving the best possible result for the level of investment (not neglecting equity), Equity is about ensuring that benefits are distributed fairly. As we can see, these objectives may be conflicting, and care must be taken to prevent optimization conflicts. Cost-effectiveness emphasizes the ultimate impact of the policy, program or project relative to the inputs invested. VfM is a key factor to consider when planning policies, programmes and projects and when taking any decisions involving the use of public resources.

Several methods can be used to assess VfM [29]: Cost-Effectiveness Analysis (CE analysis), Cost-Utility Analysis (CU analysis), Cost–Benefit Analysis (CB analysis), Social Return on Investment (SROI), Rank correlation of cost vs impact, and Basic Efficiency Resource Analysis (BER analysis). Some of these methods are common to generic economic evaluation. The primary purpose of economic evaluation is to inform decisions, so it deals with both inputs and outputs (costs and consequences) of alternative courses of action and is concerned with choices. The main types of economic evaluation studies are cost analysis (without identification or measurement of consequences), cost-effectiveness analysis, cost-utility analysis, and cost–benefit analysis [19].

Figure 7.1 combines the Santana et al. [72] framework for implementing PCC with the VfM framework for different assessment types considering all the value cycle from inputs to impact. This model thus allows a broad governance perspective of implementing PCC at various decision-making levels (macro, meso and micro). It also reinforces the importance of value from measuring health outcomes that matter to patients against the cost of delivering the outcomes. Finally, focus on the importance of obtaining the maximum benefit with the resources available proving to the different stakeholders (citizens, founders, policymakers, and managers) that the additional health care resources needed to make the procedure, service, or program available to those who could benefit from it are justified. It is essential to mention that this broad perspective of governance should be guided by the principles of good governance commonly identified in the literature, namely strategic vision, responsiveness, effectiveness and efficiency, accountability, transparency, equity and the rule of law [42, 84, 86]. The accountability of decision-makers (in government, the private sector and civil society organizations) to the public and stakeholders, and the transparency of processes, institutions and information, directly accessible to those concerned with them, and providing enough information to understand and monitor, are two core fundamental principles.

Any evaluation framework must also consider that the measurement system embraces two functions. The first is the act of the measurement itself, in other words, a technical activity consisting of "the assignment of numbers to observations" [14]. The second is the psychological effect the numbers exert on the employees who are responsible for the measured activities. Thus, measurement systems could be understood as "psycho-technical systems" [28].

Looking at the measurement system from the perspective of PCC implementation and functioning, medical organizations achieve their goals, when it delivers medical services better satisfying patients' needs with greater efficiency and effectiveness than by "usual care". Here the term effectiveness refers to the extent to which patients' requirements are met, while efficiency is a measure of how economically the provider or healthcare system resources are utilized when providing a given level of patients well-being. To achieve the above goals, organizations have to influence employees in such a way that they will work according to the PCC framework. Although, organizations will provide appropriate training and resources they have to implement a measurement system to control whether established goals are achieved.

Generally, two elements can be measured, either directly the results of human work (e.g. how many gears were made by the miller during the working day) or human behavior (e.g. compliance of the procedure performed by the doctor with medical guidelines) [59]. Which element should be measured depends on the characteristics of the work performed. If the measurement of work performance can be precise and the task can be programmed, that is, there are no significant deviations during the work process, that means, the tasks are routine, then the measurement could be conducted based on an assessment of behavior or the final result of the work. On the other hand, when tasks become less programmable, that means, more stochastic, many non-standard situations may arise that require autonomous decisions based on many variables. In such a case, only the final result of the performed task can be used as the measurement strategy. For example, the car showroom owner is not interested in what actions the salesforce took and what techniques they use. What matters is how many cars they sold and at what price.

If the measurement of results is ambiguous, and the tasks are programmable, then the only available measurement strategy is the evaluation of behavior. For example, receptionists in a hospital. They have strictly defined routine tasks to perform, and the results of their work depend little on their actions, but on how many patients appear on a given day. However, what is important is their behavior, which means, whether they are not leaving their workplace, are polite, helpful and know the applicable procedures. It is important to state that behaviour is just one piece in the measurement puzzle because to have better organizations/systems we also need good decisions and information search and processing.

The problem appears, when both the measurement of results is ambiguous and the tasks are complex and non-programmable, then the measurement is rather impossible. In this case, employees' behavior could be influenced through the implementation of social mechanisms, such as goal alignment, selection and socialization of employees. For example, the work of a team of surgeons in a trauma centre. In this action, the necessary behavior cannot be determined in advance, since it is not known what injuries the patient will suffer from. It is also impossible to determine how many patients a day they have to operate.

7.3 The Components of Value on PCC Implementations—Metrics, Outcomes and Costs

We need robust outcomes measurement approaches and costs determination to use the different instruments provided from VBHC, VfM and economic evaluation in PCC settings. Continually improving value requires better tools to assess both costs and benefits in health care [53]. These measurement systems are essential not only for decision support at different levels (macro, meso and micro), for evaluation (and inform decisions), and for management. From a management perspective, what is not measured cannot be managed or improved and, consequently, linking cost to process improvements or outcomes is difficult [65]. Using metrics to measure PCC can help drive the changes needed to improve the quality of healthcare that is person-centred [69].

Several authors recognize the difficulties associated with metrics and measures on PCC outcomes. Silva [77], states that there are many tools available to measure PCC, but no agreement about which tools most worthwhile and that currently there are no standardized mechanisms to measure and monitor PCC at a healthcare system level. In the same line, Pendrill [62] argues that there is no "silver bullet" or best measures covering all aspects of PCC. From a metrological point of view, there is a need to go beyond the traditional physical, chemical and biological quantities of clinical biomarkers to develop reference standards for novel kinds of quantities related to patient activity, participation and social well-being. Santana et al. [69] also point out that international efforts are being made towards the PCC model, but there are no standard mechanisms at a healthcare system level to measure and monitor PCC. The main conclusion is that a measuring instrument can be the best in one context and not so good in another, and there is a continuous need to ensure that quality indicators (often called psychometric indicators) are the most appropriate for the specific case.

There are several examples of international efforts to develop standard mechanisms. The OECD [58] recognizes as a meaningful action to reorient health systems to become more people-centred the need for investment in measures that help to assess whether health systems deliver what matters to people, and not only how much they cost. To fill in this gap, this international organization received a mandate from Health Ministers, to launch the Patient-Reported Indicators Surveys (PaRIS) initiative in 2017, to benchmark outcomes that matter most to patients (EIT [20]. Several other organizations, like the International Consortium for Health Outcomes Measurement (ICHOM), the Patient-Reported Outcomes Measurement Information System (PROMIS), the Outcome Measures in Rheumatology (OMERACT), and the Consensus-based Standards for the selection of health Measurement (COSMIN), The Health Foundation, and the Person-Centred Care Team also work on this field.[2]

Metrics and measurement

The purpose of the measurement system is to influence human behavior to achieve organizational goals. The exploration of the literature suggests that there are four modes through which the act of measuring can influence people's behavior at work [27, 28]. The first mode refers to the goals setting and their measures of achievement. In this mode, the measurement system serves as a criterion operationally defining the goals and standards of behavior or levels of targets (results) that should be achieved. In the context of PCC, this can relate to the establishment of targets of outcome or standards of behavior concerning the three genetic routines: initiating a partnership, shared decision-making; and safeguarding the partnership by documenting the person's narrative and the jointly agreed care plan. The establishment of standards and targets simplifies the reality and, in this way, may have robust psychological effects. In other words, these standards and targets may provide a model of an appropriate set of variables to which actions should be adjusted and therefore focuses employees' efforts and helps to organize thoughts and directions of analyzes.

The second mode is to mobilize managers and employees to systematic planning. Regular measurement of employees' activity according to simplified criteria (predefined indicators or standards), catalyzes to increase both managers and employees planning efforts to achieve these predefined indicators or standards. Flamholtz [27] claims that measurement forces managers to think systematically about human resources value. They must anticipate future requirements for the workforce and their needs for training and development concerning the targets and standards they must accomplish. Additionally, managers have to assess the value of these tasks and standards to the organization's success. In these circumstances, the numbers produced may not be as significant as the measurement process itself. This may suggest that measuring subjective constructs such as human resource value or compliance with PCC routines may not be a critical limitation. Although the measured numbers may be uncertain, the measurement process may strengthen systematic planning [27]. In this mode, the informative function of the measurement system may be regarded as a form of ex-post control as it allows the results achieved to be compared with the plan. On the other hand, when the measurement is used to indicate the expected level of achievement, it performs a planning-like function, influencing the behavior

---

[2] A more detailed table will be presented in the topic "Health outcomes that matter to patients". While the table contains specific examples for some areas, like rheumatology, other examples exist for other clinical specialties.

of employees while pursuing the assumed goals, therefore it can be considered an ex-ante control.

The third mode of the measurement system is to influence the perception of employees. The measurement system creates a set of information that serves as input and that generates alternatives for decision making and problem-solving. Therefore, decision and behavioral alternatives could be limited to the set of information generated by the measurement system.

The fourth mode of the measurement system affects both the direction and magnitude of the motivation. The presence or absence of performance measurements in a particular area affects employees' behavior. Employees focus their efforts on areas where results or behavior is measured and ignore areas that are not being measured or rewarded. Measurement serves as a motivational function when information gathered about specific activities is used to evaluate an individual or group's contribution and determine rewards [27, 32].

Information about achieved targets or the extent to the adherence to standards may also serve as a control mechanism. It can control employee behavior in an organization in two ways: directional and motivational. In the first, feedback information guides behavior by providing the information necessary for corrective action. In the second, it motivates behavior by serving as a promise of future benefits. In other words, the recipient of such feedback may use this information for corrective action, may interpret it as a reward or punishment, or may interpret it as a promise of future reward or punishment. This informative feedback function seems most appropriate in the context of a control system of employees. Extant studies suggest that the impact of feedback on results is positive when information is frequent, from a reliable source, timely, comprehensible, task-relevant, and specific [28].

Health outcomes that matter to patients

Several authors identify different types of outcome measures and tools that can be used in PCC settings. From a VfM and economic evaluation perspective, Gold et al. [31], Sorenson et al. [80], and Drummond et al. [19] identify clinical outcomes, quality of life measures, and generic health gain measures like Quality-adjusted life years (QALYs), the Disability-Adjusted Life-Year (DALY), SF-36, EQ5D, and SF6D. The ECHOs model (Economic, Clinical, and Humanistic Outcomes) focus on economic outcomes and their interrelationships with the clinical and humanistic outcomes [13]. Other studies identify tools that can be applied to different actors and dimensions [46, 77] and quality indicators that can be used to measure PCC [67–69]. The explored framework from Santana et al. [71], presented in the previous section, considers as outcomes PROMs, PREMs, and PRAOs. We extended that framework by considering PRIs. Table 7.1 presents several sources of health outcomes, tools and instruments that can be used to evaluate PCC implementations.

Clinical outcomes are the most common health outcome category to be considered in clinical trials and observational studies. Some economic evaluations use these outcomes to measure health gain [19]. The humanistic outcomes are based on the patient perspective (e.g. reported scales that indicate pain level, degree of

Table 7.1 Sources of health outcomes, tools and instruments

International Consortium for Health Outcomes Measurement (ICHOM)
https://www.ichom.org I ICHOM collaborates with patients and healthcare professionals to define and measure patient-reported outcomes to improve care quality and value. Are made available several standardized outcome measurement tools, time points, and risk adjustment factors for specific given conditions are available
Patient-Reported Outcomes Measurement Information System (PROMIS)
www.healthmeasures.net I PROMIS includes measures (over 300) of physical, mental, and social health for the general population and individuals living with chronic conditions
Outcome Measures in Rheumatology (OMERACT)
omeract.org I OMERACT is an independent initiative of international stakeholders interested in outcome measurement
Consensus-based Standards for the selection of health Measurement (COSMIN)
www.cosmin.nl I COSMIN aims to improve the selection of outcome measurement instruments in research and clinical practice by developing methodology and practical tools for selecting the most suitable outcome measurement instruments
Patient-Reported Indicators Surveys (PaRIS)
www.oecd.org/health/paris I PaRIS is an OECD's initiative where countries work together to develop, standardise, and implement a new generation of indicators that measure the outcomes and experiences of health care that matter most to people
The Health Foundation
www.health.org.uk I The Health Foundation is an independent charity that links their knowledge from working with healthcare providers and their research and analysis
Measuring patient experience—approaches to measuring patient and carer experiences of healthcare (www.health.org.uk/publications/measuring-patient-experience)
Helping measure person-centred care—a review of evidence about commonly used approaches and tools to measure person-centred care (www.health.org.uk/publications/helping-measure-per son-centred-care)
Person-Centred Care Team
https://www.personcentredcareteam.com I The Person-Centred Care Team measure person-centred care and the patient experience to improve it while partnering with the community in health research
PC-QIs—Monography of Person-Centred Care Quality Indicators (https://www.personcentre dcareteam.com/s/PC-QIs_Monograph_Santana-et-al-2019.pdf)

functioning) and include health-related quality of life (HRQoL) and the range of measures collectively described as patient-reported outcomes (PROs).

PROMs, are information provided by the patient about their symptoms, quality of life, adherence, or overall satisfaction [49, 85]. Data is collected by generic and disease-specific validated tools related to the quality of life (e.g. EQ-5D, AqoL), symptoms (e.g. NPRS for pain, FSS for fatigue), distress (e.g. K10 or PHQ2 for depression, GAD7 for anxiety), functional ability (e.g. WHODAS 2.0, ODI), self-reported health status (e.g. SF-36), or self-efficacy (e.g. GSE). Those tools collect patient ratings about several outcomes (health status, health-related quality of life, symptoms, functioning, satisfaction with care, or treatment satisfaction) or reports about their health behaviours, including adherence and well-being habits.

PREMs are tools and instruments that report patient satisfaction scores with health service [85] that are often used to capture health care's overall patient experience. They are often used on the broader population and in non-specific settings such as an outpatient department, emergency services, or inpatient services. They are a reliable measure of how well health organizations can provide good quality service from a patient perspective on several dimensions: waiting time, access and ability to navigate services, involvement in decision-making, knowledge of the care plan and pathways, quality of communication, support to manage a long-term condition, or if they would recommend the service to family and friends.

PRAOs are related to adverse events and adverse outcomes [8]. The adverse events may be caused by all the aspects of care, including ameliorable adverse events (injuries whose severity could have been substantially reduced if different actions or procedures had been performed or followed) and preventable adverse events (injuries that could have been potentially avoided). The adverse outcome is any suboptimal outcome experienced by the patient, including a new or worsening symptom, an unexpected visit to a health facility, or death that can be identified by medical record review, discharge summaries, or through a patient interview. If the adverse outcome is captured latter, it is defined as PRAO. According to Barbara et al. [8], PRAO is any adverse outcome reported directly by the patient without interpretation of anyone else. Knowing patients' perspectives and their relatives about quality and safety have become a priority, helping to build care processes centred on their users and improve clinical teams and organizations [56]. So, the perspective of PRAOs goes beyond the boundaries of health organizations and can be used in several fields like to improve the predictive accuracy of clinicians [9] or evaluate discharge communication for preventing death or hospital readmission [70].

PRIs are a more recent categorization of data proposed by Baldwin et al. [6] that upgrades the PRO tool reinforcing the patient's perspective. This categorization considers the growing importance of social networks and how patients publish and receive communications easily and quickly. Social media has been used increasingly and more frequently by diverse stakeholders, including patients, to find information about diseases and treatments, and o to give support to patients and their caregivers. Schlesinger et al. [73] identified four forms of PRIs that can be used to improve clinical practice: patient-reported outcomes measuring self-assessed physical and mental well-being, surveys of patient experience with clinicians and staff, narrative accounts describing encounters with clinicians in patients own words; complaints/grievances signalling patients distress when treatment or outcomes fall short of expectations.

Lastly, it is essential to mention that from the previous section's perspective, where we combine two frameworks, PCC implementation and value for money, that the different measures, tools, instruments and indicators can be used from different forms considering levels of decision-making, stakeholders and categories. For instance, Silva [77], from the UK's Health Foundation, reviewed measurement tools targeted to different actors (carers, patients, professionals) and categories (carer experience, communication, dignity/empathy, engagement, patient experience, person-centred care, self-management support, and shared decision-making). From a quality perspective, the Canadian Institutes of Health Research and the MSI Foundation identified person-centred quality indicators related to structure, process, and outcomes [68, 69]. Using program theories (PTs), Lloyd et al. [46] identified seven different types of evidence-based PTs that could shape health laboratories' design to implement PCC. It is possible to identify various qualitative and quantitative measures for the measurement/assessment that could be applied to diverse stakeholders and categories.

Costs of delivering the outcomes

Like outcomes, costs and their determination are crucial for VBHC, VfM and economic evaluation. In their first approaches to VBHC, Porter and Kaplan [65] claim that there is a complete lack of understanding of how much costs deliver patient care, and how these costs compare with achieved outcomes. According to them, poor cost systems have disastrous consequences, due to the difficulty of linking cost to process improvements or outcomes, leading to huge cross-subsidies across services with distortions in the supply and efficiency of care, and supplying wrong signals to providers, not rewarding adequately the effective and efficient ones, while providing inefficient ones little incentive to improve.

The accurate cost measurement in health care is a challenging issue. From a management accounting perspective, according to Young [90], healthcare organizations and health systems face various financial challenges related to economic pressures, health care reforms, and care demand that require an understanding of the costs associated with care delivery. The main forces that affect costs are demographic changes, with more ageing people, different spending patterns for older people, increasing morbidity in the nonelderly people, mainly related to cancer and heart disease, and the healthcare market's complexity. These forces can be addressed by combining case mix and volume, resources per case, cost per resource unit, and fixed costs.

Porter and Kaplan (2011) also wrestle with the complexity of health care delivery, since treatment involves many various types of resources (human resources, equipment, technology, space and supplies) with different capabilities and costs. From admission to discharge, the patient pathway is diverse and depends on the initial medical conditions. The existence of non-standardized procedures (different procedures, drugs, devices, tests, and equipment for the same medical process) and the highly fragmented world of healthcare delivery, with different providers interventions, also contributes to this complexity.

According to Kaplan et al. [43], existing cost systems in healthcare rely on inaccurate and arbitrary cost allocation, while providing little transparency to guide first-line care providers in attempts to understand and change the proper drivers of their costs. They also argue that these systems prevent clinician-driven cost reduction and process improvement initiatives, proposing time-driven activity-based costing

Table 7.2 Some advantages and disadvantages of the TDABC

Advantages	Disadvantages
• More accurate cost estimates [12]	• Possible inaccuracy of time estimates [15, 35]
• Better use of resources, activities and processes, increasing the capacity used and eliminating those that do not add value [23, 43, 91]	• Time needed to determine time estimates [76]
• Process optimization, trying to reduce time consumed by some activities [35]	
• Best benchmarking model [76]	
• Efficiency in allocating costs to the cost object [44]	

(TDABC) as one tool with significant potential to fill this gap. However, its application to healthcare has been limited. Some of the main advantages and disadvantages of the TDABC identified in the literature are presented in Table 7.2.

From a direct economic evaluation perspective, some considerations must be taken into account. Gold et al. [31] identifies costs related to changes: in the use of healthcare resources, in the use of non-healthcare resources, in informal caregiver time and in patient time (for treatment). By its turn, Drummond et al. [19], identifies health sector costs, other sector costs, patient/family costs, and productivity losses. Table 7.3 presents the different types of costs identified by those authors.

In a general way, authors identify categories of direct costs, indirect costs and intangible costs. Direct costs associated with providing the health service (fixed, variable, and non-medical expenses) are the easiest to calculate. Indirect costs related to decreased productivity due to the patient and his family's disease or treatment are difficult to compute. Intangible costs (such as anxiety, pain or suffering with an illness) are extremely difficult or even impossible to determine.

The various perspectives presented here focus on the difficulty in determining the costs associated with healthcare interventions. If we adopt a VBHC perspective, we must considerer the full cycle of care for a particular medical condition [64, 66]. The VBHC definition applies to the entire care pathway, from primary care to tertiary care, including post-hospital care of patients affected by single and multiple conditions. From this perspective, implementing costs measurement should be relatively straightforward for patients affected by only one condition. However, it will be much more complicated for patients affected by multiple conditions and a problem to organizations with various type of services and a higher level of complexity requiring substantial investments in more accurate cost systems.

Table 7.3 Different types of costs

Gold et al. [31]
Direct health care costs
All types of resource use: consumption of professional, family, volunteer, or patient time, costs of tests, drugs, supplies, healthcare personnel, and medical facilities
Non-direct health care costs
Additional costs related to the interventions: childcare (for a parent attending a treatment), dietary prescription, transportation (to and from health facilities); time (family or volunteers spent providing home care)
Patient time costs
Time a person spends seeking care or participating in or undergoing intervention or treatment. Also include travel and waiting time as well the time receiving treatment
Drummond et al. [19]
Health sector costs
Variable (such as the time of health professionals or supplies)
Fixed or overhead costs (such as light, heat, rent, or capital costs)
Other sector costs
Consumed resources from other public agencies or the voluntary sector
Person/family costs
Any out-of-pocket expenses incurred by patients or family members
Value of any resources that they contribute to the treatment process
Productivity costs
Costs associated with lost or impaired ability to work or to engage in leisure activities due to morbidity
Lost economic productivity due to death

7.4 The Person-Centred Care Approach to Family Caregivers Needs Assessment and Support in Community Care

It should be noticed that the effective use of assessment and measurement tools depends on the application domain and their peculiarities. We will now show an application of an assessment tool (CSNAT) created to support community care for palliative care.

Community nurses play an essential role in providing palliative care for the patients, family and friends who support them at home. A unique role of providing community care to terminally ill persons mostly for several years have family members, friends, and/or neighbours. Those persons are in literature known as family caregivers or carers. Family caregivers or carers are the persons who operate home-based care with very complex medical and therapeutic tasks which are unpaid and have not formal training to provide those services. They act like an "arm extension" for providing care to terminally ill persons in the healthcare system.

Many factors hinder the proper identification and assessment of family caregiving support's need to facilitate access to appropriate resources. Assessment of family caregivers supports needs usually have an ad hoc manner, informal and undocumented without comprehensive consideration needs and family carer problems. Caregivers usually focused on the needs and issues of patients. Home-based family caregiving terminal ill patients can significantly impact emotional, social, and physical cost caregivers and increase their mortality [5, 33, 74, 81]. Studies have shown that only early identification and better carers' experience leads to better health outcomes in the longer-term, leading to reduced caregivers' morbidity and mortality [2, 26, 34].

The Carer Support Needs Assessment Tool (CSNAT) is an evidence-based tool for the comprehensive assessment of carers' needs, in a systematic way instead of an ad hoc manner [25]. This support tool for family carers[3] validated in UK CSNAT was developed to use in palliative care and adopts a screening format. The tool has a screening format structured around 14 broad support domains [25]. These domains fall into two distinct groupings, including: (1) those to enable a family carer to care at home, and (2) enable more direct support for themselves in a caring role (Table 7.4) [25]. The tool is practical brief from one side. From the other side, it is comprehensive, allowing the family to carer to identify priority domains that need further support, which can be discussed with providers (community nurse). Caregivers support cares in their primary role of caring, and their support needs will be identified by care providers.

The CSNAT is a five-step (Table 7.5) approach in person-centred care and care-led. For the appropriate use and implementation of this tool, changes must be made in existing practice, and practitioners need to shift their role from practitioner-led to practitioner-facilitated.

Benefits of the CSNAT approach

The CSNAT was used in numerous research studies and these showed benefits for practitioners and carers. The CSNAT's was trialled in several studies in Western Australia [1, 2, 4] achieving overall positive practitioners feedback on using the CSNAT. Similar results have been obtained in a study that aims to assess the feasibility and relevance of CSNAT for home-based care at family caregivers of people with motor neurone disease (MND) [3]. This study was conducted in Perth (Western Australia). The family caregivers and advisers found that CSNAT approach is relevant and acceptable for the patient with MND in home-based care. This study also confirmed previous studies that used CSNAT approach about the comprehensive and formalized compared to standard practice. It was the first example of applying the CSNSAT approach in MND settings, which was different from the approach developed in the United Kingdom, and from the further trials in Australia in home-based

---

[3] This tool is oriented to family carers. In the field of informal care, the carer is not always a patient's relative, however we kept the tool's nomenclature.

Table 7.4 Support domains of the Carer Support Needs Assessment Tool (CSNAT). *Source* Ewing et al. [24]

Seven domains of support enabling the carer to care (co-worker role)	
Carer identifies whether he/she needs more support with:	Understanding their relative's illness
	Managing their relative's symptoms, including giving medicines
	Providing personal care (e.g. dressing, washing, toileting)
	Knowing whom to contact when concerned
	Equipment to help care for their relative
	Talking with their relative
	Knowing what to expect in the future when caring for their relative
Seven domains of support concerning their own wellbeing (co-client role)	
Carer identifies whether he/she needs more support with:	Looking after he/she own physical health
	Having time for oneself in the day
	Any financial, legal, or work issues
	Dealing with feeling and worries
	Beliefs or spiritual concerns
	Practical help in the home
	Getting a break from caring overnight

Table 7.5 Steps in the CSNAT approach. *Source* Horseman et al. [39]

Step	Explanation
Introduction of the CSNAT	The practitioner administers the CSNAT by introducing and explaining it at the earliest opportunity in the caregiving journey
Caregivers' consideration of needs	The practitioner allows time for the caregivers to consider their needs using the CSNAT
Assessment conversation	An assessment conversation takes place, in which the caregiver highlights their support need priorities
Shared action plan	The assessment conversation leads to development and documentation of an action plan, which summarises actions required from the caregiver and practitioner
Shared review	Regular review of the caregivers' needs

palliative care settings. These caregivers identified and addressed three of five priorities to support, mainly priorities related to direct carer support such as knowing what to expect in the future, dealing with feelings and worries and having time for themself. In this study, for the caregivers, high priority was expressed for the knowledge of whom to contact if he/she feels concerned and needs care equipment.

A stepped-wedge cluster-randomized trial was conducted in Australia, which aimed to investigate the impact of using the CSNAT on family caregivers' outcomes such as strain, distress and mental and physical health, and describing investigation strategies [1]. This study showed that the mean (0.08) reduction in caregiver strain in the intervention group compared to the control group increased by 0.09 after adjusting for covariates. Means difference between an investigation and compared groups related to distress were not statistically significant. There was a small difference in secondary outcome (mental and physical wellbeing), but those differences were not statistically significant among groups. In both groups showed an increase in caregivers workload assisting with daily activities of daily living, but this increase was smaller and was not statistically significant. In the implementation strategy, caregivers put on the top four support needs, such as knowing what can expect in the future, having time for themselves in the day, dealing with their feelings and worries, and understanding the relatives' illness.

A trial with family caregivers of older people discharged home from the hospital showed that family caregivers in the intervention group were significantly more prepared to provide care and reported reduced carer strain and distress than family caregivers in the control group [83].

Barriers and facilitates in implementing of the CSNAT approach

Implementation of CSNAT represents a challenge for the practitioners. Implementing it can block due to various barriers: practitioners' beliefs and attitudes, lack of knowledge or training, and lack of time or resources [39]. The barrier can also be the number of practitioners ready to adopt the implementation of the new tool. According to the evidence, the number of adopters grows with a high proportion of internal facilitators among the staff members [16]. Training and giving the authority to the internal team facilitator is also an essential factor in the implementation process's success, as is recognizing the crucial role of carers in providing home-based care for terminally ill patients, which also need to be enabled and supported by the clinical, managers, educators, and policymakers. In terms of health system governance, another barrier is the lack of visibility of this full-time work in society and the usual lack of recognition by the government initiatives and policies. Recommendations regarding the CSNAT approach suggested that the tool should form the basis of carer needs assessment, rather than an add-on to current practice [11].

7.5   Conclusion

When dealing with PCC interventions, it is necessary to have a global view of the entire system considering, from a health governance perspective, the different levels of decision making, the multiple stakeholders and the alignment of their interests. As mentioned before, although PCC innovations are implemented at the micro-level by care providers, they also are implemented at the organizational level (meso) and at a higher level (macro) by structural decisions that affect the shape of the entire health system. We also need to evaluate the impact of these innovative models on the whole society. VBHC, VfM and economic evaluation provide concepts, methodologies, and tools that can be used to compare costs and consequences evaluating their impact. Still, we need accurate outcomes and costs measurement systems.

The effectiveness of the measurement system's influence on the behaviour compliant with the PCC routines may depend mainly on the adequacy and reliability of the information generated by the measurement system. In this context, adequacy of a measurement system refers to the extent to which the measurement process leads to the employees' behavior supporting the PCC routines. While reliability refers to the extent to which the behaviours formed by the measurement process are consistently repeated. That means, that in many circumstances regardless of other factors, the measurement process motivates employees to follow the PCC approach.

As an important facilitator for practitioners in implantation, an evaluation tool should provide education on evidence-based knowledge, and be incorporated in an organizational environment which supports organizational learning and interaction in exchanging knowledge and experience about implantation.

Acknowledgements This publication is based upon work from COST Action "**European Network for cost containment and improved quality of health care-CostCares**" (CA15222), supported by COST (European Cooperation in Science and Technology)

COST (European Cooperation in Science and Technology) is a funding agency for research and innovation networks. Our Actions help connect research initiatives across Europe and enable scientists to grow their ideas by sharing them with their peers. This boosts their research, career and innovation.

https://www.cost.eu

References

1. Aoun, S., Deas, K., Toye, C., Ewing, G., Grande, G., Stajduhar, K.: Supporting family caregivers to identify their own needs in end-of-life care: qualitative findings from a stepped wedge cluster trial. Palliat. Med. **29**, 508–517 (2015)
2. Aoun, S., Toye, C., Deas, K., Howting, D., Ewing, G., Grande, G., Stajduhar, K.: Enabling a family caregiver-led assessment of support needs in home-based palliative care: potential translation into practice. Palliat. Med. **29**, 929–938 (2015)
3. Aoun, S.M., Deas, K., Kristjanson, L.J., Kissane, D.W.: Identifying and addressing the support needs of family caregivers of people with motor neurone disease using the Carer Support Needs Assessment Tool. Palliat. Support. Care **15**, 32–43 (2017)
4. Aoun, S.M., Grande, G., Howting, D., Deas, K., Toye, C., Troeung, L., Stajduhar, K., Ewing, G.: The impact of the Carer Support Needs Assessment Tool (CSNAT) in community palliative care using a stepped wedge cluster trial. PLoS One **10,** e0123012 (2015)
5. Aoun, S.M., Kristjanson, L.J., Currow, D.C., Hudson, P.L.: Caregiving for the terminally ill: at what cost? Palliat. Med. **19**, 551–555 (2005)

6. Baldwin, M., Spong, A., Doward, L., Gnanasakthy, A.: Patient-reported outcomes, patient-reported information. Patient **4**, 11–17 (2011)
7. Banke-Thomas, A., Nieuwenhuis, S., Ologun, A., Mortimore, G., Mpakateni, M.: Embedding value-for-money in practice: a case study of a health pooled fund programme implemented in conflict-affected South Sudan. Eval. Program Plann. **77**, 101725 (2019)
8. Barbara, O., Jose, S.M., Jayna, H.-L., Ward, F., Maeve, O.B., Deborah, W., Wrochelle, O., Ghali, W.A., Forster, A.J.: A framework to assess patient-reported adverse outcomes arising during hospitalization. BMC Health Serv. Res. **16**, 357 (2016)
9. Basch, E., Bennett, A., Pietanza, M.C.: Use of patient-reported outcomes to improve the predictive accuracy of clinician-reported adverse events. J. Natl. Cancer Inst. **103**, 1808–1810 (2011)
10. Bhattacharyya, O., Blumenthal, D., Stoddard, R., Mansell, L., Mossman, K., Schneider, E.C.: Redesigning care: adapting new improvement methods to achieve person-centred care. BMJ Qual. Saf. **28**, 242–248 (2019)
11. Carer Support Needs Assessment Tool.: The CSNAT Approach: a person-centred process of carer assessment and support in palliative and end of life care (2016). Available from: http://csnat.org/files/2015/10/161111_The-CSNAT-Approach_general-use.pdf
12. Chen, A., Sabharwal, S., Akhtar, K., Makaram, N., Gupte, C.M.: Time-driven activity based costing of total knee replacement surgery at a London teaching hospital. Knee **22**, 640–645 (2015)
13. Cheng, Y., Raisch, D.W., Borrego, M.E., Gupchup, G.V.: Economic, clinical, and humanistic outcomes (ECHOs) of pharmaceutical care services for minority patients: a literature review. Res. Social Adm. Pharm. **9**, 311–329 (2013)
14. Cohen, B.P.: Developing Sociological Knowledge: Theory and Method. Wadsworth Publishing Company (1989)
15. Crott, R., Lawson, G., Nollevaux, M.-C., Castiaux, A., Krug, B.: Comprehensive cost analysis of sentinel node biopsy in solid head and neck tumors using a time-driven activity-based costing approach. Eur. Arch. Otorhinolaryngol. **273**, 2621–2628 (2016)
16. Diffin, J., Ewing, G., Harvey, G., Grande, G.: The Influence of context and practitioner attitudes on implementation of person-centered assessment and support for family carers within palliative care. Worldviews Evid. Based Nurs. **15**, 377–385 (2018)
17. Dodgson, R., Lee, K., Drager, N.: Gobal Health Governance: A Conceptual Review. Dept of Health & Development World Health Organization, Centre on Global Change & Health London School of Hygiene & Tropical Medicine (2002)
18. Donabedian, A.: The quality of care: how can it be assessed? JAMA **260**, 1743–1748 (1988)
19. Drummond, M.F., Sculpher, M.J., Claxton, K., Stoddart, G.L., Torrance, G.W.: Methods for the Economic Evaluation of Health Care Programmes. Oxford University Press (2015)
20. EIT Health: Implementing Value-Based Health Care in Europe: Handbook for Pioneers (2020)
21. Ekman, I., Busse, R., Hoof, C.V., Klink, A., Kremer, J.A., Miraldo, M., Olauson, A., Rosen-Zvi, M., Smith, P., Swedberg, K., Törnell, J.: Healthcare innovations and improvements in a financially constrained environment - Strategy Plan and R&D Roadmap (WE Care Consortium). WE Care Consortium (2016)
22. Ekman, I., Wolf, A., Olsson, L.-E., Taft, C., Dudas, K., Schaufelberger, M., Swedberg, K.: Effects of person-centred care in patients with chronic heart failure: the PCC-HF study. Eur. Heart J. **33**, 1112–1119 (2012)
23. El Alaoui, S., Lindefors, N.: Combining time-driven activity-based costing with clinical outcome in cost effectiveness analysis to measure value in treatment of depression. PLoS One **11**, e0165389 (2016)
24. Ewing, G., Austin, L., Diffin, J., Grande, G.: Developing a person-centred approach to carer assessment and support. Br. J. Community Nurs. **20**, 580–584 (2015)
25. Ewing, G., Brundle, C., Payne, S., Grande, G.: The Carer Support Needs Assessment Tool (CSNAT) for use in palliative and end-of-life care at home: a validation study. J. Pain Symptom Manage. **46**, 395–405 (2013)

26. Ferrario, S.R., Cardillo, V., Vicario, F., Balzarini, E., Zotti, A.M.: Advanced cancer at home: caregiving and bereavement. Palliat. Med. **18**, 129–136 (2004)
27. Flamholtz, E.G.: Toward a psycho-technical systems paradigm of organizational measurement. Decis. Sci. **10**, 71–84 (1979)
28. Flamholtz, E.G., Das, T.K., Tsui, A.S.: Toward an integrative framework of organizational control. Acc. Organ. Soc. **10**, 35–50 (1985)
29. Fleming, F.: Evaluation methods for assessing Value for Money (2013). Available from: http://betterevaluation.org/sites/default/files/Evaluating%20methods%20for%20a ssessing%20VfM%20-%20Farida%20Fleming.pdf
30. Fors, A., Taft, C., Ulin, K., Ekman, I.: Person-centred care improves self-efficacy to control symptoms after acute coronary syndrome: a randomized controlled trial. Eur. J. Cardiovasc. Nurs. **15**, 186–194 (2015)
31. Gold, M.R., Siegel, J.E., Russell, L.B., Weinstein, M.C., Russell, L.B.: Cost-effectiveness in Health and Medicine. Oxford University Press (1996)
32. Goncharuk, A.G., Lewandowski, R., Cirella, G.T.: Motivators for medical staff with a high gap in healthcare efficiency: comparative research from Poland and Ukraine. Int. J. Health Plann. Manage. **35**, 1314–1334 (2020)
33. Grande, G., Stajduhar, K., Aoun, S., Toye, C., Funk, L., Addington-Hall, J., Payne, S., Todd, C.: Supporting lay carers in end of life care: current gaps and future priorities. Palliat. Med. **23**, 339–344 (2009)
34. Grande, G.E., Farquhar, M.C., Barclay, S.I.G., Todd, C.J.: Caregiver bereavement outcome: relationship with hospice at home, satisfaction with care, and home death. J. Palliat. Care **20**, 69–77 (2004)
35. Gregório, J., Russo, G., Lapão, L.V.: Pharmaceutical services cost analysis using time-driven activity-based costing: a contribution to improve community pharmacies' management. Res. Social Adm. Pharm. **12**, 475–485 (2016)
36. Groenewoud, A.S., Westert, G.P., Kremer, J.A.M.: Value based competition in health care's ethical drawbacks and the need for a values-driven approach. BMC Health Serv. Res. **19**, 256 (2019)
37. Health Innovation Network: What is person-centred care and why is it important? (2014)
38. Hernandez, S.E., Conrad, D.A., Marcus-Smith, M.S., Reed, P., Watts, C.: Patient-centered innovation in health care organizations: a conceptual framework and case study application. Health Care Manage. Rev. **38**, 166–175 (2013)
39. Horseman, Z., Milton, L., Finucane, A.: Barriers and facilitators to implementing the Carer Support Needs Assessment Tool in a community palliative care setting. Br. J. Community Nurs. **24**, 284–290 (2019)
40. ICAI: ICAI's Approach to Effectiveness and Value for Money. Independent Commission for Aid Impact (ICAI) (2011)
41. ICAI: DFID's approach to value for money in programme and portfolio management: a performance review (2018)
42. Institute on Governance: Governance Basis - What is governance? Available from: https://iog. ca/what-is-governance/ (2021). Accessed 2021–02–10 2005
43. Kaplan, R.S., Witkowski, M., Abbott, M., Guzman, A.B., Higgins, L.D., Meara, J.G., Padden, E., Shah, A.S., Waters, P., Weidemeier, M., Wertheimer, S., Feeley, T.W.: Using time-driven activity-based costing to identify value improvement opportunities in healthcare. J. Healthcare Manage. **59**, 399–412 (2014)
44. Keel, G., Savage, C., Rafiq, M., Mazzocato, P.: Time-driven activity-based costing in health care: a systematic review of the literature. Health Policy **121**, 755–763 (2017)
45. Larsson, S., Lawyer, P.: Improving Health Care Value: The Case for Disease Registries. The Boston Consulting Group (2011)
46. Lloyd, H.M., Ekman, I., Rogers, H.L., Raposo, V., Melo, P., Marinkovic, V.D., Buttigieg, S.C., Srulovici, E., Lewandowski, R.A., Britten, N.: Supporting innovative person-centred care in financially constrained environments: the WE CARE exploratory health laboratory evaluation strategy. Int. J. Environ. Res. Public Health **17**, 3050 (2020)

47. Lorenzoni, L., Murtin, F., Springare, L.-S., Auraaen, A., Daniel, F.: Which policies increase value for money in health care? (2018)
48. Louw, J.M., Marcus, T.S., Hugo, J.F.: How to measure person-centred practice-An analysis of reviews of the literature. Afr. J. Primary Health Care Family Med. **12**, 1–8 (2020)
49. Mackinnon III, G.E.: Understanding Health Outcomes and Pharmacoeconomics. Jones & Bartlett Publishers (2011)
50. McCormack, B., McCance, T.: Development of a framework for person-centred nursing. J. Adv. Nurs. **56**, 472–479 (2006)
51. Mccormack, B., Mccance, T.: Person-centred nursing: theory and practice. Wiley –Blackwell (2010)
52. McCormack, B., McCance, T.: Person-centred Practice in Nursing and Health Care: Theory and Practice. John Wiley & Sons (2016)
53. McGinnis, J.M., Olsen, L., Yong, P.L.: Value in Health Care: Accounting for Cost, Quality, Safety, Outcomes, and Innovation: Workshop Summary. National Academies Press (2010)
54. Mjåset, C., Ikram, U., Nagra, N.S., Feeley, T.W.: Value-Based Health Care in Four Different Health Care Systems. NEJM Catalyst Innovations in Care Delivery, 1 (2020)
55. Nilsson, K., Bååthe, F., Andersson, A.E., Wikström, E., Sandoff, M.: Experiences from implementing value-based healthcare at a Swedish University Hospital - an longitudinal interview study. BMC Health Serv. Res. **17**, 169–169 (2017)
56. O'Hara, J.K., Reynolds, C., Moore, S., Armitage, G., Sheard, L., Marsh, C., Watt, I., Wright, J., Lawton, R.: What can patients tell us about the quality and safety of hospital care? Findings from a UK multicentre survey study. BMJ Qual. Saf. **27**, 673 (2018)
57. OECD: Health Care Systems: Getting More Value for Money. Author Paris, France (2010)
58. OECD: Ministerial Statement - The Next Generation of Health Reforms: Ministerial Statement. OECD (2017)
59. Ouchi, W.G.: A Conceptual framework for the design of organizational control mechanisms. Manage. Sci. **25**, 833–848 (1979)
60. Peek, C., Higgins, I., Milson-Hawke, S., McMillan, M., Harper, D.: Towards innovation: the development of a person-centred model of care for older people in acute care. Contemp. Nurse **26**, 164–176 (2007)
61. Pendleton, R.: We won't get value-based health care until we agree on what "value" means. Harvard Bus. Rev. **2**, 2–5 (2018)
62. Pendrill, L.R.: Assuring measurement quality in person-centred healthcare. Meas. Sci. Technol. **29**, 034003 (2018)
63. Phelan, A., Mccormack, B., Dewing, J., Brown, D., Cardiff, S., Cook, N.F., Dickson, C., Kmetec, S., Lorber, M., Magowan, R.: Review of developments in person-centred healthcare. Int. Pract. Dev. J. (2020)
64. Porter, M.E.: What is value in health care?. N. Engl. J. Med. **363**, 2477–2481 (2010)
65. Porter, M.E., Kaplan, R.S.: How to solve the cost crisis in health care. Harvard Bus. Rev. **4**, 47–64 (2011)
66. Porter, M.E., Kaplan, R.S.: How to pay for health care. Harvard Bus. Rev. **94**, 88–100 (2016)
67. Santana, M.-J., Ahmed, S., Lorenzetti, D., Jolley, R.J., Manalili, K., Zelinsky, S., Quan, H., Lu, M.: Measuring patient-centred system performance: a scoping review of patient-centred care quality indicators. BMJ Open **9**, e023596 (2019)
68. Santana, M.-J., Manalili, K., Ahmed, S., Zelinsky, S., Quan, H., Sawatzky, R.: Person-centred Care Quality Indicators (2019). Available from: https://www.personcentredcareteam.com/s/PC-QIs_Monograph_Santana-et-al-2019.pdf
69. Santana, M.-J., Manalili, K., Zelinsky, S., Brien, S., Gibbons, E., King, J., Frank, L., Wallström, S., Fairie, P., Leeb, K., Quan, H., Sawatzky, R.: Improving the quality of person-centred healthcare from the patient perspective: development of person-centred quality indicators. BMJ Open **10**, e037323 (2020)
70. Santana, M.J., Holroyd-Leduc, J., Southern, D.A., Flemons, W.W., O'Beirne, M., Hill, M.D., Forster, A.J., White, D.E., Ghali, W.A.: A randomised controlled trial assessing the efficacy of an electronic discharge communication tool for preventing death or hospital readmission. BMJ Qual. Saf. **26**, 993 (2017)

71. Santana, M.J., Manalili, K., Jolley, R.J., Zelinsky, S., Quan, H., Lu, M.: How to practice person-centred care: a conceptual framework. Health Expectations **21**, 429–440 (2018)
72. Santana, M.J., Manalili, K., Jolley, R.J., Zelinsky, S., Quan, H., Lu, M.: How to practice person-centred care: a conceptual framework. Health Expectations **21**, 429–440 (2018)
73. Schlesinger, M., Grob, R., Shaller, D.: Using patient-reported information to improve clinical practice. Health Serv. Res. **50**(Suppl 2), 2116–2154 (2015)
74. Schulz, R., Beach, S.R.: Caregiving as a risk factor for mortality the caregiver health effects study. JAMA **282**, 2215–2219 (1999)
75. Sharma, T., Bamford, M., Dodman, D.: Person-centred care: an overview of reviews. Contemp. Nurse **51**, 107–120 (2015)
76. Siguenza-Guzman, L., Auquilla, A., van den Abbeele, A., Cattrysse, D.: Using time-driven activity-based costing to identify best practices in academic libraries. J. Acad. Librariansh. **42**, 232–246 (2016)
77. Silva, D.D.: Helping Measure Person-Centred Care - A Review of Evidence About Commonly Used Approaches and Tools Used to Help Measure Person-Centred Care. The Health Foundation (2014)
78. Smith, P.C.: Measuring Value for Money in Healthcare: Concepts and Tools. The Health Foundation, London (2009)
79. Soderland, N., Kent, J., Lawyer, P., Larsson, S.: Progress Towards Value-Based Health Care. Lessons from 12 Countries. The Boston Consulting Group. Inc. (2012)
80. Sorenson, C., Drummond, M., Kanavos, P.: Ensuring Value for Money in Health Care: The Role of Health Technology Assessment in the European Union. WHO Regional Office Europe (2008)
81. Stajduhar, K.I., Funk, L., Toye, C., Grande, G.E., Aoun, S., Todd, C.J.: Part 1: home-based family caregiving at the end of life: a comprehensive review of published quantitative research (1998–2008). Palliat. Med. **24**, 573–593 (2010)
82. The Economist Intelligence Unit: Value-based Healthcare: A Global Assessment Findings and Methodology (2016)
83. Toye, C., Parsons, R., Slatyer, S., Aoun, S.M., Moorin, R., Osseiran-Moisson, R., Hill, K.D.: Outcomes for family carers of a nurse-delivered hospital discharge intervention for older people (the Further Enabling Care at Home Program): Single blind randomised controlled trial. Int. J. Nurs. Stud. **64**, 32–41 (2016)
84. UNDP: Reconceptualising Governance. New York: Management Development and Governance Division Bureau for Policy and Programme Support - United Nations Development Programme (1997)
85. Verma, R.: Overview: What are PROMs and PREMs. NSW Agency for Clinical Innovation (2016)
86. WHO: Good Governance for Health. Equity Initiative Paper No 14. Geneva: Department of Health Systems, World Health Organization (1998)
87. WHO: People at the Centre of Health care: Harmonizing Mind and Body, People and Systems. WHO Regional Office for the Western Pacific, Manila (2007)
88. WHO: Primary Health Care Now More Than Ever. World Health Report. Geneva, Switzerland: World Health Organization (2008)
89. WHO: WHO Global Strategy on People-centred and Integrated Health Services (2015)
90. Young, D.W.: Management Accounting in Health Care Organizations. Jossey-Bass (2014)
91. Yu, Y.R., Abbas, P.I., Smith, C.M., Carberry, K.E., Ren, H., Patel, B., Nuchtern, J.G., Lopez, M.E.: Time-driven activity-based costing to identify opportunities for cost reduction in pediatric appendectomy. J. Pediatr. Surg. **51**, 1962–1966 (2016)

Chapter 8
Studying the Impact of Human Resources on the Efficiency of Healthcare Systems and Person-Centred Care

Bojana Knezevic, Roman Andrzej Lewandowski, Anatoliy Goncharuk, and Maja Vajagic

Abstract We explore the alternative explanation for barriers and facilitators for implementation of PCC evolving from human resources through the lenses of institutional theory. We have deepened the explanation by adding the perspective of different institutional logics, which shows that the physician's resistance or nurses' support may originate from the differences in institutional logics. Working with patients by applying person-centered principles places new demands on health professionals. It is widely agreed that education and training are very important for the clarification on the roles of professionals in the person-centred care. PCC education programs were designed to be delivered through informal training, continued medical education, leadership development and training through mentors' system. Managers, on the other hand, may support the implementation of PCC, but their motivation may be less oriented to increase of service quality, than gaining higher external legitimacy of the organization and increase organizational access to external resources. Therefore, managers may not implement sufficient control and motivational mechanisms for healthcare professionals for following PCC routines and make them slip back into 'usual care' or lose interest, knowledge or commitment. As the psychological state of medical staff can determine the duration and success of the treatment and care,

B. Knezevic (✉)
University Hospital Centre Zagreb, Zagreb, Croatia
e-mail: bojana.knezevic@kbc-zagreb.hr

R. A. Lewandowski
Institute of Management and Quality Science, Faculty of Economics, University of Warmia and Mazury, 10-720 Olsztyn, Poland
e-mail: rlewando@wp.pl

A. Goncharuk
International Humanitarian University, Odessa, Ukraine
e-mail: agg@ua.fm

M. Vajagic
Croatian Health Insurance Fund, Zagreb, Croatia
e-mail: maja.vajagic@hzzo.hr

R. A. Lewandowski
Voivodeship Re-Habilitation Hospital for Children in Ameryka, Olsztynek, Poland

© The Author(s) 2022
D. Kriksciuniene and V. Sakalauskas (eds.), *Intelligent Systems for Sustainable Person-Centered Healthcare*, Intelligent Systems Reference Library 205,
https://doi.org/10.1007/978-3-030-79353-1_8

therefore they should be properly motivated. In this chapter we show comparative research study in Ukraine and Poland. The methodology of this study selected a list of motivators for medical staff in both countries. The results of this study brought the main findings that may be useful for reforming inefficient healthcare systems.

Keywords Human resources · PCC education · Hospital managers · Institutional theory · Motivation in healthcare

8.1 Introduction

Human resources in health care system are the different kinds of clinical and non-clinical staff responsible for public and individual health intervention. Human resources are one of the most important inputs leading to the efficient output of health care system [39]. In addition to doctors and nurses, there are many other professionals involved in treatment and care processes, such as psychologists, laboratory staff, pharmacists, dietitians, and others. For many centuries health care was managed by medical professionals, however, rising costs of health care services forced governments' to impose tighter controls over the medical practice by the implementation of New Public Management (NPM) reform. As a result, professional managers with their business-like logic has been introduced in medical settings previously dominated by medics with their professional, care logic [45]. However, the implementation of NPM together with managers has not restrained the increase of expenditure for medical services and safeguard quality improvement [41]. Thus, a new way of cost containment and quality improvement has to be found.

The problem is extremely important since in the near future the European healthcare systems could be financially unsustainable and fail to protect the current level of access and quality of medical services. The main problems are demographic changes as the aging population has resulted in people living longer, but also a growing prevalence of chronic long-standing illnesses [77]. The rising number of people with complex care needs requires the development of care systems that bring together a range of professionals and skills from the healthcare, long term and social care sectors [24]. Long-term diseases are today the leading cause of mortality worldwide and are estimated to be the leading cause of disability [23]. In this context quality improvement in healthcare should consist of systematic and continuous actions that lead to a measurable enhancement in healthcare services [36]. The most important goals for improvement in healthcare are safe, effective, patient-centred, timely, efficient, and equitable healthcare [37]. The quality improvement and cost containment are challenging since the healthcare sector is highly fragmented, organized along with different subsectors, disciplines and diseases [13].

Recently, promising results concerning quality improvements and cost containments have been achieved by implementing Person-Centred Care (PCC) in medical services (Fors et al. 2017). Person-centred care is today widely advocated as a key component of effective illness management. PCC has been shown to advance the match between a provider of care and patient on treatment plans, improve health outcomes and increase patient satisfaction [23]. PCC is a way of thinking and doing things that sees the people using health and social services as equal partners in planning, developing and monitoring care to make sure it meets their needs. This means putting people and their families at the centre of decisions and seeing them as experts, working alongside professionals to get the best outcome [59].

From the research appears, however, that many implementations of PCC are difficult to become sustainable [4]. Early research has recognized that the whole context of the care environment and system of education could be the source of potential barriers [51, 66]. Studies looking in more detail in the care context identified that professional practice, beliefs and cultures are the most prominent obstacles [8]. More recent research, to some extent, confirmed previously identified barriers and defined them as: traditional practices and structures, skeptical, stereotypical attitudes from professionals; and factors related to the development of person-centred interventions [56]. Moore et al. [56] also identify facilitators for PCC implementation, which are closely related to the barriers. They claim that organizational factors, professionals' attitude, leadership and training as well as the way PCC was delivered across projects may facilitate its implementation. Although McCormack [8], Bolster and Manias [51] and Moore et al. [56] identified barriers and facilitators for the implementation and functioning of PCC, they have not clearly explained what mechanisms make these barriers and facilitators arise. Thus in this chapter, we tried to fill this scientific gap by looking at these barriers and facilitators from the institutional theory point of view.

Although initial studies indicate the positive influence of PCC on quality improvement and cost containment (Fors et al. 2017), the results to some extent may be biased. In the literature analyzing PCC implementation [56, 75] authors have recognized important problems but have not explained their sources. In our opinion, these problems may be the symptoms of structural issues concerning such phenomena as institutional decoupling, rivalry between different medical professionals such as nurses and physicians as well as coexistence of various institutional logics such as professional logic and managerial logic [44]. Problems recognized by Moore et al. [56] and Naldemirci et al. [58] may also originate from the fact that PCC to some extent could function as a "powerful myth" [54], this means that PCC could not be implemented in order to increase quality and contain costs but to gaining external social legitimacy. In this chapter, we have not intended to say that PCC does not work. But the inference about the above mentioned structural problems and recognition that PCC may be implemented as a myth may have far-reaching consequences. The most important corollaries may include unrealistic perception of PCC effects, the extent to which PCC is recommended for various treatment and care interventions, and how human resource management should be organized during PCC implementation and functioning. In this chapter, we deepen the understanding of the role of healthcare and its institutional logics in the context of PCC implementation. We also describe the role of the motivation of healthcare workers in the process of changing their way of thinking.

8.2   Person-Centred Care in the Context of Human Resource Management in Healthcare

In everyday reality and practice, disease-focused and clinician centred care are emphasized treating a disease, without attention to the needs of the patient, and centred on the health professional as the sole source of control. Working with patients in a more person-centred manner places new demands on health professionals [66]. Person-centred care means treating patients as individuals and as equal partners in the process of healing. Person-centred care supports people to develop the knowledge, skills and confidence they need to make informed decisions about their own health [35]. It is not a medical model and should be regarded as a multidisciplinary approach, recognizing that a person may need more than one professional to support them. The multidisciplinary approach in hospitals is most often dominated by medical professionals [36]. Health care professionals who are involved in the healing process could be consist of doctors, nurses, psychologists, social workers, pharmacists, dietitians, managers in healthcare and others. Most people want to help themselves, so the health system should be able to ensure that they acquire the knowledge, skills and confidence to do in this way [15] Working in the way of person-centred care means recognizing people's capabilities and potential to manage and improve their own health, not seeing them simply passive recipients of care or as victims of the disease. Professionals often underestimate the extent to which patients are able to take responsibility for their health. Many patients would be willing and indeed eager to do so if their capabilities were recognized, supported and strengthened [15]. For those people who are with limited or without the mental capacity to assume greater responsibility for their care, the system should ensure family, relatives, home caregivers and trained advocates who have to be fully involved in the care planning process.

Person-centred care (PCC) is a new way of thinking and doing things that see people using health and social services as equal partners. Patients, medical professionals and healthcare managers, as well as policymakers should work together and develop structures to measure and monitor PCC performance and to promote PCC practice for suitable interventions [66]. As well, PCC has to be monitored based on feedback from patients since they should be the ultimate reviewer of the efficiency of PCC functioning. Objective measurement of whether PCC brings quality improvement and cost containment (see Chap. 3) is extremely important since as we have proved further—the implementation and functioning of PCC could be distorted. Mostly due to complex and multiparadigmatic human resource relationships.

8.3   Studying the Human Resources Impact on PCC

From the human resource and the organizational theory perspective, organizations are the collection of individuals that could be structured according to many criteria, for example, as members of the organizational unit, specific professional group and also different informal social networks existing within an organization. However, from the institutional theory, we could perceive human resources as a collection of organizational actors with different organizational logics decoupling their "usual practices" ensuring organizational effectiveness from ceremonial conformity with externally legitimate rules.

In order to explain causes and sources of barriers to PCC implementation and functioning we did not perform our own field research, instead, we based our inference on previous studies that tried to explain Alharbi et al. [4], Moore et al. [56] and Naldemirci et al. [58] observations in more detail and from the institutional theory perspective.

8.3.1 PCC Implementation as External Pressure

Naldemirci et al. [58], based on normalization process theory (NPT) [50] tried to discover the way organizations implement PCC. They revealed that organizations use two strategies to implement PCC: deliberate and emergent. They claimed that deliberate strategies consist of training, seminars and financing, and proved to be necessary and effective in disseminating the knowledge about PCC but were not sufficient to guarantee positive implementation outcomes. Emergent strategies, involved informal meetings, discussing real situations, 'reflecting on action' and 'learning in action'. Although the authors used the label of "emergent strategies" we do not find evidence in the research about the pure bottom-up initiative [58]. As in the literature, there is no clear cut-off between deliberate and emergent strategies we accept the label "emergent strategies" in the sense that "the central leadership intentionally creates the conditions under which strategies can emerge" [55]. This entitles us to the conclusion that PCC implementations as describe them Moore et al. [56] and Naldemirci et al. [58] took place rather by outside pressure than through bottom-up processes leading by professionals working directly with patients. Therefore, all analyses described further in the chapter were conducted under the assumption that the implementation of PCC was initiated by the constituency external to the organization (scientific project, local government programs) or external, at least, to the professionals working directly with patients (managerial initiatives). To clarify our scientific argument we adopt a simplification that the purpose of PCC implementation is to change the 'usual care' practices into PCC routines.

8.3.2 PCC as a Powerful Myth

Although there is evidence that PCC is an effective tool for quality improvement and cost containment (e.g., Fors et al. 2017), and gained significant public support especially in Sweden, for example, through government funding of the GPCC, it could not be treated as a universal solution. It means that in some settings externally driven PCC implementation may not much to the clinical demands and type of patients. The lack of fit of PCC routines to performed medical services in the organization could not be recognized as a definite premise against PCC implementation but rather as some barriers that must be overcome. If the PCC would be implemented in bottom-up processes by professionals directly working with patients, then the fit of PCC routines to clinical demands and type of patients would be highly probable.

However, in externally driven implementation, explained in the previous section, the degree of fit is unknown. Then, the fit could range from the perfect fit to the misfit. But it might be difficult to recognize in advance, before the implementation, at what point of the continuum clinical demands and type of patients, treated in a medical organization are located. The recognition of the fit is problematic since in externally driven PCC implementation the initial resistance to this change cannot be treated as the objective sign of the misfit. Thus, it is unknown to what extent the pressure should be executed to overcome the barriers to the PCC implementation. Where is the point when the training and experience gained by the time of implementation convince health workers and PCC routines will be normalized, what means "it becomes routinely embedded in the matrices of already existing, socially patterned, knowledge and practices" [50].

Even though it might be visible before the implementation that PCC routines do not fit certain clinical demands and types of patients in an organization, reviling this by professionals and managers from the organization may be difficult. The more PCC had gained recognition and support from political bodies there would be higher social pressure to implement PCC in every medical organization. PCC has already being promoted by many scholars and powerful organizations, such as WHO and the European Committee for Standardization (CEN)[1] as a "taken-for-granted solution" capable of improving quality and contain costs in almost every medical organization. Such kind of external legitimacy has a high potential to establish PCC as an institutional rule (norm) thus, enforce institutional isomorphism [20]. In this context, PCC could be also perceived by many organizations as a "powerful myth" in a sense this notion is described in the seminal work of [54].

8.3.3 Implications for PCC Implementation as a Powerful Myth

The inference that PCC could be perceived as an institutional rule functioning as a powerful myth has many important implications and allow shedding new light on the barriers of implementation of PCC reviled in previous studies. In this environment, a ceremonial adaptation of PCC would be the natural reaction to protect the efficiency

---

[1] CEN published The European Standard EN 17,398:2020, "Patient involvement in health care—Minimum requirements for person-centred care" (https://iteh.fr/catalog/tc/cen/bc1d2237-3a90-46e8-a976-89c0ae53c7cc/cen-tc-450).

of "usual care" against the external implementation of new routines. From the literature, we know that to resolve the conflict between ceremonial rules and efficiency organizations call for decoupling [54]. Decupling means that under external pressure an organization creates two worlds. The one visible from the outside, consisting of symbolic activity acceptable by external constituencies and therefore giving an organization legitimacy and access to resources. And the second, internal, following "usual" practices that allow the organization to render efficiently their day-to-day work [45].

Decoupling may have a tremendous influence on researchers' inference investigating the results of PCC implementation. Scholars, unaware of the possibility of decoupling, may perceive ceremonial conformity decoupled from everyday practices only as some minor lack of adherence to PCC routines instead of deep institutional problems. For example, scientists revealing that "Several researchers described working with professionals who said they were practicing PCC when they were not." Moore et al. [56] should dig further into the field research to discover whether the PCC routines were followed only as a ceremonial activity to "show off" in front of the external spectator (e.g., the scientist) or it was only "conflicting and/or divergent views about PCC and the difficulty of translating abstract principles into concrete practices." [58, 56]. In this situation, the difference between interpretations of the results of the investigation could be totally opposite. In the first case, a researcher should conclude that the PCC was not implemented successfully and probably it would be difficult to find any solution to improve the results of the implementation. In the second case, a scholar could assume that the implementation was quite successful, only some employees need more training, motivation and experience with PCC.

The decoupling, however, could be deeply embedded and during the research may not be discovered. Thus, a researcher needs more premises to interpret some lack of compliance with PCC routines as a "ceremonial implementation". From the literature we know, that organizations with easily measured outcomes and high potential to prove objectively their contribution to the society (e.g. surgical wards) may resist openly the implementation of PCC and do not need to go into ceremonial activities [54]. As an example, we can bring Moore et al. [56] study in which they reported that "The surgical setting proved a particularly tough climate for PCC because of high patient turnover and standardized prescribing". Hence, in this case, a researcher could be more certain about the results of their investigation knowing that organizational actors were more honest.

Organizations providing services that are difficult to assess are more willing to incorporate institutionalized rules bringing external trust and confidence to their outputs [54]. Therefore, the decision about the implementation of PCC could be perceived with higher suspicion as an attempt to increase organizational legitimacy. These types of organizations in order to expand their survival prospects may implement PCC regardless of sharp conflict with current efficiency criteria and well-suited practices and procedures. Scholars studying these organizations should take into consideration the potential ceremonial implementation of PCC.

8.3.4 Institutional Logics

Building on the above discussion we can further deepen our understanding of causes and sources of implementation barriers of PCC through using another perspective—institutional logics. From the perspective of institutional theory, day-to-day practices of organizational actors are the result of institutional logics existing in an organization since institutional logics are "socially constructed, historical patterns of material practices, assumptions, values, beliefs and rules" [74], "...the basis of taken-for-granted rules" ([62], p. 629) and "frames of reference that condition actors' choices for sense-making, the vocabulary they use to motivate action, and their sense of self and identity" [76]. Institutional logics deliver a belief system and associated practices and are the organizing principles that shape cognition and behavior [47, 62, 67, 71, 75]. An organization is characterized by institutional complexity, embracing multiple logics, as opposed to being dominated by a single logic. These different logics influence organizational strategy, structure and especially day-to-day practices [32].

These multiple logics may cohabitate in different configurations, they could be competitive, co-operative, orthogonal or blurred [30]. However, relationships between logics might be subject to interpretation by individual organizational actors as they can execute some degree of agency and selectively employ, interpret, and enact logics or some parts of them [38, 53]. The coexistence of multiple logics is especially complex in organizations with high interdependency, as in healthcare, where at least three main logics cohabitate: physicians institutional logic with core value focusing on the diagnosis, inference and treatment [1], nurses institutional logic concentrating on care and managers logic with core values relating to economic and business issues. These logics do not differ only in values, rules and practices but they also vary in hierarchical relationships and realms of responsibility.

8.3.5 Influence of Nurses and Physicians Institutional Logics on PCC Implementation

Physicians, for example, always have exercised a high degree of autonomy and clinical judgment, remain relatively free from external regulations and reluctantly transfer their core work to other professions, using an arsenal of methods to defend their territory [1, 25, 26]. Nurses on the other hand align with more holistic care for the patient compared to physicians [17]. "The role of nursing [...] is to care and that of a doctor is to cure." [52]. The situation of nurses is also significantly different in terms of hierarchical relationships in the workplace. Although nursing during the last decades went through a process of professionalization, changed the status from personnel rendering simple auxiliary functions under direct physicians' supervision to professionals offered medical advice [72] is still perceived as inferior to physicians [52]. Managers, in a sense of top organizational leaders, exercise the highest freedom and driving force in medical settings. In contrast to the previous two professions, managers values and area of activity, focuses not on patients but on the whole organization. Taking into account the above discrepancies between institutional logics, it is unlikely that these three intuitional logics may respond similarly to the implementation of PCC.

From Alharbi et al. [4], Moore et al. [56] and Naldemirci et al. [58] studies appear that doctors, resisted the implementation of PCC, while nurses were much more supportive of implementing these new practices. This difference could be explained by two causes. The first explanation relates to hierarchical interdependence. Physicians, as was mentioned earlier, have had always high status and superior position, and were not interested in any change. These cannot be said about nurses. Nurses, therefore, had seen the change, the implementation of PCC as an opportunity to improve their position in the organizational hierarchy. They expressed this in the interview: "In person-centered care nurses get a new role, actually, in the team, because we have our individual task to perform (…). But we have a different role and we need to work together with the physician, not under the physician" [56, 75]. Nurses, and other "auxiliary" professionals support for organizational change such as PCC implementation is not unique. A similar situation was observed, for example, during the healthcare reform in Canada. There, registered nurses and physiotherapists supported the change since they realized that the new system 'could result in a higher profile for their profession' [63].

The second explanation relates to the significance and visibility of input each profession delivers to the final output of the medical organization. Professions similarly to organizations have to prove their effectiveness and value of contributions to the outcome to maintain or improve their external legitimization in the society [26]. Physicians with their well-defined tasks consisting of diagnosis, inference and treatment [1] did not feel the need to change their practice to gain more legitimacy from PCC implementation, which might be perceived by researchers as resistance. Nurses, on the other hand, performing in many situations auxiliary tasks, again could see the implementation of PCC as an opportunity for gaining more external legitimization as an independent profession.

8.3.6 Managers Institutional Logic Influence on PCC Implementation

Since the 1980s, after the introduction of the New Public Management (NPM) paradigm to health care systems, managers have started to play important role in medical organizations. As a result, additional managerial logic has been introduced to organizations earlier dominated by professional logics [44]. Thus, in medical settings managers have become another factor that had to be taken into account during the implementation of any important change, such as the implementation of

PCC. Managerial role in the PCC implementation has been also acknowledged by previous research:

"Another barrier was the inter-professional hierarchy between doctors and nurses. The implementation of this framework depended upon better cooperation between different professional groups. However many early adopters came from the nursing profession and some doctors were not initially keen to embrace the model. Existing hierarchies posed problems for the collective action and support of managers was required" [58].

The excerpt from the interview has shown that managers played a facilitating role in the implementation process. Managers also support the implementation of PCC, for example by devoting resources in the form of relieving some nurses from their duties to became PCC "ambassadors" [58]. But the managerial motivation to implement PCC might be different than those of nurses. Nurses may perceive practices related to PCC as the vehicle supporting their claims to higher organizational and professional positions, hence they were eagerly rendering PCC routines. Managers responsible for ensuring organizational access to external resources might be rater motivated by the increase of external legitimacy of the organization.

Some studies raise the issue that PCC is more time and resource consuming than "usual care" and lower organizational productivity which may lead to a decrease of efficiency (e.g.: [4, 75. It did not mean that PCC failed in cost containment but the savings may be not within an organization but outside—other parts of the healthcare system. This is the case when PCC leads to reduced "time of hospitalization" [58]. But the reduced time of hospitalization, may increase patients' turnover and consequently workload for the ward staff and consumption of hospital resources. Such a situation should discourage managers from PCC implementation. Since from the literature, we know that in some hospitals, managers use even ethically questionable methods to increase efficiency. For example, some managers use case-mix (DRG) tariffs to screen patients and push pressure on physicians to admit only those patients who are "profitable" [14, 43] and use specialist software which advises, and also controls physicians whether they apply optimal lengths of stay and proper medical procedures to patients concerning their conditions (diagnostic codes) to maximize revenues from the payer [65]. Taking into account potentially contradictory managerial activities, on the one side supporting the implementation of PCC which could lower organizational economic efficiency and on the other side, strong actions to increase the economic efficiency—deeper analysis is needed. The problem with the perception of PCC as a "powerful myth" is that managers to increase legitimacy not only conform to myths but also maintain the appearance that the myths actually work [54].

This contradiction can be explained in terms of institutional theory. As it was already mentioned above, managers may resolve the contradictory activities by decoupling, in other words, separating employees' internal day-to-day activities from the externally visible routines. This would mean that the implementation of PCC could be to some extent ceremonial, especially in areas where the resources are the most expensive. Ceremonial implementation of PCC could be explained by two phenomena observed by Alharbi et al. [4], Moore et al. [56] and Naldemirci et al. [58]. First, during the field study, the scholars observed "professionals who said they were practicing PCC when they were not." [56] and most of them were physicians [58] who are the most expensive part of human resources. Second, that researchers after returning to the medical setting some time after the implementation observed a significant rate of abandonment of PCC routines [4]. The abandonment could mean that organizations after gaining external visibility as PCC-practicing, slept back to their "usual care".

In the above sections, we explain, based on institutional theory, some causes and sources of barriers created by human resources. Further, we describe facilitators supporting PCC implementation.

8.4   Education and Training Programs in PCC

International experts agree that education and training are very important for the clarification on the roles of professionals in person-centred care. With the evolution of PCC, there is a need for innovative education programs that are endorsed by key stakeholders, including medical faculty, administrative directors and accrediting bodies. Educational programs should also include administrative staff, volunteers and other professionals involved in care, who are needed to support the cultural change. As integrating PCC into the health care curriculum does not directly lead to implementation into practice, PCC education programs should be designed to continue through informal training, continued medical education, continued leadership development and training through mentors system [66

Healthcare workers who choose to work in person-centred care places new demands on them. Current education tends to focus on the biomedical model, and it is not developed with patients and all health-care providers. Future models of education incorporate both perspectives in the development of training. It requires some specific skills as excellent listening, communication and negotiation skills and the capacity to respond flexibly to people's individual needs [66, 15].

Institute of Medicine (IOM) conclude two decades before that all health professionals should be educated to deliver patient-centred care as members of an interdisciplinary team, emphasizing evidence-based practice, quality improvement approaches, and informatics [37].

Modified the conclusion of IOM, the most important topics in PPC education and training are:

- **Provide person-centred care**—identify, respect, and care about patients as a person. Recognize every person (professionals and private persons) who take care of ill person. Take into a count personal difference, values, preferences, and expressed needs; relieve pain and suffering; coordinate continuous care; listen to, clearly inform, communicate with, and educate patients; share decision-making and management; and continuously advocate disease prevention, wellness, and promotion of healthy lifestyles.

- **Work in interdisciplinary teams**—cooperate, collaborate, communicate, and integrate care in teams to ensure that care is continuous and reliable. Different means of experts and settings for delivering care, such as managed care, community-based care, rehabilitation centres, and critical pathway systems, require interdisciplinary teams to provide the necessary coordination.
- **Evidence-based-medicine**—integrates the best research with clinical expertise and patient values for optimum care, and participates in learning and research activities to the extent feasible.
- **Quality improvement in healthcare**—identify risk in care; understand and implement basic safety design principles, such as standardization and simplification; continually understand and measure the quality of care in terms of structure, process, and outcomes in relation to patient and community needs; and design and test interventions to change processes and systems of care, with the objective of improving quality.
- **Use new technologies and informatics**—use of telemedicine, Web-based communication channels, using patients preferred communication channels, manage knowledge, mitigate error, and support decision making using information technology.

Effective evidence-based medicine is central to what most patients need, but person-centred care cannot be reduced to guidelines. Guidelines and protocols are important, but they must not exclude the important human qualities of caring and compassion, which are highly valued by patients Human resources in health sector changes also seek to improve the quality of services and patients' satisfaction.

At the organization level, management and payment systems should encourage healthcare providers to espouse the values of people-centred care. Clinical governance and other quality improvement initiatives can be established to monitor and improve provider behavior [34].

The health care financing system which largely does not reimburse professionals for time spent coordinating and integrating care or providing care through alternative vehicles, such as over the Internet or via telephone-further constrains clinicians' efforts to care for patients [37].

Quality improvement leaders in hospitals need to be included in the development of these programs through the measure for improving the quality of the work process. One of the barriers to PCC implementation in healthcare is the result of the lack of emphasis on PCC in medical education. The lack of understanding of person-centred care concept could lead to a lack of motivation for implementation [66].

8.5 Study of Health Professionals' Motivation and Efficiency of Healthcare Systems

We rarely think about how our health is not so much in our own hands but it is in the hands of the medical staff that treats and cares for us. Respectively, it is obscure to contemplate the difficulties these people face in terms of mood, workplace stress and satisfaction with their careers as well as overall lives—especially in a time of need. The psychological state of medical staff, thus, can determine the duration and success of the treatment. To improve this state—by making it adequate and positive—doctors, nurses and other health professionals should be properly motivated, i.e., induced for high efficiency and quality of work.

Obviously, the healthcare systems, in which medical staff is properly motivated, show a higher level of effectiveness and efficiency and, vice versa, improperly motivated a lower level [3, 21, 61, 64, 70, 73]. Recent research on healthcare performance in Europe applied data envelopment analysis and revealed countries with high and low effectiveness and efficiency rates continent-wide [46]. Generally, healthcare effectiveness is measured as the ratio of healthy years lived and life expectancy per infant mortality [3, 70, 73]. However, according to Lo Storto and Goncharuk's approach [46], the effectiveness of the healthcare system is based solely on the quality of care. Besides, the healthcare efficiency reflects how many medical doctors, nurses, and other health professionals in correlation with available beds in hospitals are needed to meet the demand for healthcare services among the population of a country. They stated the higher the number of medical staff and hospital beds versus demand interlinked with an overall higher level of healthcare efficiency. Moreover, during the period from 2011 to 2014, only three countries remained continuously the most efficient healthcare systems in Europe, namely, Ireland, Portugal and Poland. It should be stressed, however, that Sweden jumped by almost 50% and became relatively effective in 2014 [46].

Recently Del Rocío Moreno-Enguix et al. [18] also identified the leading countries across the continent and found the United Kingdom, Cyprus and Poland as the most efficient. Moreover, other prominent studies have rated the Polish system as the highest efficient healthcare system in Europe [9, 27, 42]. Utilizing this standpoint, Poland remains the best country to assess healthcare efficiency with different approaches to efficiency assessment [10]—leading us to look deeper into its experience in terms of medical staff motivation. As such, learning about how and what motivates Polish medical staff to show high efficiency underscores an important premise in a medical and European context.

On the other side, Lo Storto and Goncharuk [46] showed that the healthcare system of Ukraine was the least efficient among 33 European countries. Goncharuk [28] studied the motivation of Ukrainian medical staff in terms of essential features from an external and internal perspective, including different motivators for various groups of medical staff (i.e., profession, gender and age) as well as significant disparities between Ukrainian health professionals and their colleagues from other countries. These differences may explain the performance of Ukrainian medical staff and the inefficiency of its healthcare system.

As a result, Goncharuk et al. [29] assumed the behaviour and motivation of medical staff in a low-performing healthcare system (i.e. Ukraine) is fundamentally different from an efficient one (i.e. Poland). The authors have compiled research to test comparative findings for different motivators of medical staff in countries with a high gap in healthcare efficiency by examining the motivation of health professionals in Polish and two Ukrainian hospitals.

Despite the motivators and incentives necessary for improving employee performance and organisational effectiveness [2, 11, 22, 48, 49, 57, 60, 68], we have not found in-depth studies on the relationship between medical staff motivation and healthcare efficiency (i.e. at the national level). So we fulfilled this gap by testing different motivators for medical staff in countries with a high gap in healthcare efficiency.

The methodology of this study selected a list of motivators for medical staff in countries with a high gap in healthcare efficiency—Polish, i.e., considered efficient and Ukrainian, i.e. considered inefficient. The study tested the following null hypothesis H0: there is not a significant difference between the motivators for medical staff of the two countries with a high gap in healthcare performance. So, in the event that H0 is true, we state that the motivation of medical staff has not a significant relationship with the efficiency of the healthcare system. We used a null hypothesis significance test to verify the zero difference [69]. The study employed the following six-stage scheme illustrated in Fig. 8.1.

In this study, we utilised the "Evaluation of motivators questionnaire for medical staff" adapted from Goncharuk's [28] groundwork on assessing medical staff incentives in Ukraine. A twenty-six-question questionnaire was put together using a Likert scale (i.e., 1—very unsatisfied, 2—unsatisfied, 3—neutral, 4—satisfied and 5—very satisfied) and pilot tested with medical staff of Odessa hospitals in Ukraine in the last quarter of 2018.

After a thorough fine-tuning of the questionnaire, the hospitals in Poland as well as in Ukraine agreed to partake in the questionnaire's experimentation. In stage two, we selected medical staff in Poland and Ukraine. In stage three, we conducted the questionnaire in the two countries in the first quarter of 2019. In the absence of the possibility of interviewing all Polish and Ukrainian health professionals, we have

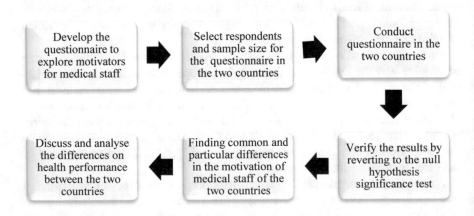

Fig. 8.1 Six-stage scheme of the study

formed two national samples, which include respondent groups that correspond in composition (i.e. profession, gender and age) to the framework of medical staff in these countries.

We considered three of the main groups of respondents in every country: experienced doctors, nurses and intern doctors. A total of 268 respondents participated in the survey, i.e., 142 respondents in Poland and 126 in Ukraine.

Descriptive statistics for the two national samples are presented in Table 8.1.

In stage four, we verified the results by reverting to the null hypothesis significance test, by using the Kruskal–Wallis test by ranks [40], a non-parametric method for testing whether samples originate from the same distribution. In stage five, if the null hypothesis is not confirmed, i.e. H0 = false, an analysis of the differences in motivation between Polish and Ukrainian responses is conducted—testing each motivator. Statistical analysis incorporated the well-known Pearson correlation coefficient [6] with the Chaddock scale [12] as well as one-way analysis of variance (ANOVA) [5, 16]. And finally, in stage six, we discuss these differences in the context of possible positive or negative influences for medical staff motivation in both countries. Using this information, we can recommend healthcare reformers and policymakers the appropriate steps to improve healthcare performance via the assessed motivators.

Table 8.1 Descriptive statistics for national samples of medical staff

Country	Variable	Groups	Total	Mean[†]	Median[†]	Stand. dev.[†]
Poland	Profession	Experienced doctors	46	20	21	1.5
(n = 142)		Nurses	70			
		Intern doctors	26			
	Gender	Male	21	46	43	15.6
		Female	121	43	44	13.4
	Age	18–34	28	43	44	1.6
		35–44	49			
		45–54	34			
		55+	31			
Ukraine	Profession	Experienced doctors	14	17	15	1.2
(n = 126)		Nurses	34			
		Intern doctors	78			
	Gender	Male	25	25	23	6.7
		Female	99	25	21	10.8
	Age	18–34	108	39	37	1.3
		35–44	9			
		45–54	5			
		55+	4			

[†]Profession = based on years of experience, gender and age = based on mean age

The results of this study brought the main findings that may be useful for reforming inefficient healthcare systems, namely:

(1)   Working conditions in an inefficient healthcare system are perceived by medical staff much worse than in an efficient one. However, medical staff in an inefficient system can be more optimistic.
(2)   Medical staff in efficient and inefficient healthcare systems has different influencing motivators.
(3)   There are huge differences in the motivation of medical staff in terms of profession, gender, and age. Hence, to achieve a high-performance healthcare system, this difference must be considered when developing a system of incentives for medical staff.
(4)   The number of motivators for experienced doctors and nurses in an inefficient healthcare system is much less than in an efficient one; furthermore, with older age, this difference becomes higher. These findings are probably due to the disappointing fact that medical staff working in a system that has poor conditions lack proper facilities.
(5)   Medical staff in an efficient healthcare system is well-motivated by moral, internal and external (i.e., financial) incentives. In an inefficient system, experienced medical staff is motivated mainly by moral incentives, but with older age, the need for social benefits increases especially for female staff.

This study has brought us closer to discovering a possible influence on the motivation of medical staff on the efficiency of healthcare systems [9, 27, 46]. Further research will need to establish the presence (or absence) of correlative means between medical staff motivation and healthcare system efficiency. As with the comparative research from Poland and Ukraine, the identified significant differences in motivation, in terms of efficient versus inefficient system, is refuted leaving the longing sense that such a relationship may be possible. Overall, the research gives us hope that interlinkages can be further established—utilizing the five recommended reform enablers.

8.6   Conclusion

In this Chapter we discussed human resources impact, and alternative explanation for barriers and facilitators for implementation of PCC evolving from human resources through the lenses of institutional theory. Managers may support the implementation of PCC, but their motivation would be not only to increase service quality. Quality improvement leaders in hospitals need to be included in the development of these programs through the measure for improving the quality of the work process. One of the barriers to PCC implementation in healthcare is the result of the lack of PCC education. The lack of understanding of person-centred care concept could lead to a lack of motivation for implementation The authors of research in Ukraine and Poland have compiled research to test comparative findings for different motivators of medical staff in countries with a high gap in healthcare efficiency by examining the motivation of health professionals in Polish and two Ukrainian hospitals. The results of this study brought the main findings that may be useful for reforming inefficient healthcare systems.

Acknowledgements 1 This publication is based upon work from COST Action "**European Network for cost containment and improved quality of health care-CostCares**" (CA15222), supported by COST (European Cooperation in Science and Technology)

COST (European Cooperation in Science and Technology) is a funding agency for research and innovation networks. Our Actions help connect research initiatives across Europe and enable scientists to grow their ideas by sharing them with their peers. This boosts their research, career and innovation.

https://www.cost.eu

Acknowledgements 2 This work was supported by the National Science Centre Poland (Grant Number: 2015/17/B/HS4/02747).

References

1. Abbott, A.: The System of Professions: An Essay on the Division of Labor. University of Chicago Press (1988)
2. Aduo-Adjei, K., et al.: The impact of motivation on the work performance of health workers (Korle Bu Teaching Hospital): evidence from Ghana. Hosp. Practices Res. 1(2), 47–52 (2016)
3. Afolabi, A., et al.: The effect of organisational factors in motivating healthcare employees: a systematic review. Br. J. Healthc. Manag. 24(12), 603–610 (2018)
4. Alharbi, T.S., et al.: Implementation of Person-centred Care: Management Perspective. JHA 3(3), 107 (2014). https://doi.org/10.5430/jha.v3n3p107
5. Annerstedt, M., et al.: Inducing physiological stress recovery with sounds of nature in a virtual reality forest—Results from a pilot study. Physiol. Behav. 118, 240–250 (2013)
6. Benesty, J., et al.: Pearson correlation coefficient. In: Cohen, I., Huang, Y., Chen, J., Benesty, J. (eds.) Noise Reduction in Speech Processing, Springer, Berlin, Heidelberg, pp. 1–4 (2009)
7. Berghout, M., et al.: Healthcare professionals' views on patient-centered care in hospitals. BHC Health Serv. Res. 15, 385 (2015). https://doi.org/10.1186/s12913-015-1049-z
8. Bolster, D., Manias, E.: Person-centred interactions between nurses and patients during medication activities in an acute hospital setting: Qualitative observation and interview study. Int. J. Nurs. Stud. 47(2), 154–165 (2010). https://doi.org/10.1016/j.ijnurstu.2009.05.021
9. Bortoletto, G., Favaro, D.: Healthcare efficiency across European countries during the recent economic crisis. Economia Pubblica 2019(2), 9–38 (2019)
10. Cantor, V.J.M., Poh, K,L,.: Integrated analysis of healthcare efficiency: a systematic review. J. Med. Syst. 42(1), 1–23 (2018)
11. Cerasoli, C.P., et al.: Intrinsic motivation and extrinsic incentives jointly predict performance: A 40-year meta-analysis. Psychol. Bull. 140(4), 980–1008 (2014)
12. Chaddock, R.E.: Principles and Methods of Statistics. Houghton Mifflin (1925)
13. COST Action (2017). European Network on cost containment and quality in healthcare. MoU, 2017.

14. Covaleski, M.A., et al.: An institutional theory perspective on the DRG framework, case-mix accounting systems and health-care organizations. Acc. Organ. Soc. **18**(1), 65–80 (1993). https://doi.org/10.1016/0361-3682(93)90025-2
15. Coulter, A., Oldham, J.: Person-centred care: what is it and how do we get there? Future Hosp. J. **3**(2), 114–116 (2016)
16. Cuevas, A., et al.: An ANOVA test for functional data. Comput. Stat. Data Anal. **47**(1), 111–122 (2004)
17. Currie, G., et al.: HR practices and knowledge brokering by hybrid middle managers in hospital settings: the influence of professional hierarchy. Hum. Resour. Manage. **54**(5), 793–812 (2015). https://doi.org/10.1002/hrm.21709
18. Del Rocío Moreno-Enguix et al.: Analysis and determination the efficiency of the European health systems. Int. J. Health Plann. Manage. **33**(1), 136–154 (2018)
19. Dellenborg, L., et al.: Factors that may promote the learning of person-centred care: an ethnographic study of an implementation programme for healthcare professionals in a medical emergency ward in Sweden. Adv. Health Sci. Educ. **24**, 353–381 (2019). https://doi.org/10.1007/s10459-018-09869-y
20. DiMaggio, P., Powell, W.W.: The iron cage revisited: collective rationality and institutional isomorphism in organizational fields. Am. Sociol. Rev. **48**(2), 147–160 (1983)
21. Dussault, G., Dubois, C.A.: Human resources for health policies: a critical component in health policies. Hum. Resour. Health **1**(1), 1 (2003)
22. Dwibedi, L.: Impact of employees motivation on organizational performance. Acad. Voices Multi. J. **7**, 24–30 (2017)
23. Ekman, et al.: Person-centred care-Ready for Prime Time. Eur. J. Cardiovasc. Nurs. **10**(4), iii–iii (2011). https://doi.org/10.1016/j.ejcnurse.2011.06.008
24. European Commission (2017) Tools and methodologies to assess integrated care in europe. Report by the expert group on health systems performance assessment. Luxembourg: Publications Office of the European Union
25. Freidson E (1988) Profession of Medicine: A Study of the Sociology of Applied Knowledge. University of Chicago Press
26. Freidson E (2001) Professionalism, The Third Logic: On the Practice of Knowledge. University of Chicago press
27. Goncharuk, A.G.: Socioeconomic Criteria of Healthcare Efficiency: An International Comparison. J. Appl. Manag. Investments **6**(2), 89–95 (2017)
28. Goncharuk, A.G.: Exploring a motivation of medical staff. Int. J. Health Plann. Manage. **33**(4), 1013–1023 (2018)
29. Goncharuk, A.G., et al.: Motivators for medical staff with a high gap in healthcare efficiency: comparative research from Poland and Ukraine. Int. J. Health Plann. Manage. **35**(6), 1314–1334 (2020)
30. Goodrick, E., Reay, T.: Constellations of institutional logics: changes in the professional work of pharmacists. Work. Occup. **38**(3), 372–416 (2011). https://doi.org/10.1177/0730888841140 6824
31. Greiner, A.C., Elisa ,K.: Health Professions Education. A Bridge to Quality, Institute of Medicine. Washingtone, National Academic Press (2003)
32. Greenwood, R., et al.: Institutional complexity and organizational responses. Acad. Manag. Ann. **5**(1), 317–371 (2011)
33. Harris, C., e al,: Human resource management and performance in healthcare organisations. J. Health Organ. Manag. **21**, 448–459 (2007)
34. Hazarika, I.: Health workforce governance: key to the delivery of people-centred care. Int. J. Healthc. Manage. (2019). https://doi.org/10.1080/20479700.2019.1647380
35. Health Foundation: Person-centred care made simple: What everyone should know about person-centred care. (2014). https://www.health.org.uk/publications/person-centred-care-made-simple
36. Hughes, R.G.: Patient Safety and Quality: An Evidence-Based Handbook for Nurses. Rockville: Agency for Healthcare Research and Quality (US) (2008)

37. Institute of Medicine (IOM): Crossing the Quality Chasm: A New Health System for the 21st Century. National Academy Press, Washington, D.C (2001)
38. Jones, C., Livne-Tarandach, R.: Designing a frame: rhetorical strategies of architects. J. Organ. Behav. **29**(8), 1075–1099 (2008)
39. Kabene, S.M., et al.: The importance of human resources management in healthcare: a global context. Hum. Res. Health **4**(20), (2006). https://doi.org/10.1186/1478-4491-4-20
40. Kruskal, W.H., Wallis, W.A.: Use of ranks in one-criterion variance analysis. J. Am. Stat. Assoc. **47**(260), 583–621 (1952)
41. Kuhlmann, E., et al.: "A manager in the minds of doctors:" a comparison of new modes of control in European hospitals. BMC Health Serv. Res. **13**(1), 1 (2013). https://doi.org/10.1186/1472-6963-13-246
42. Kujawska, J.: Efficiency of Healthcare Systems in European Countries-the DEA Network Approach. Metody Ilościowe w Badaniach Ekonomicznych **19**(1), 60–70 (2018)
43. Lewandowski RA (2014) Cost control of medical care in public hospitals–a comparative analysis. Int. J. Contemp. Manage. 13(1)
44. Lewandowski, R.A., Sułkowska, J.: Levels of hybridity in healthcare sector. In: Teczke, J., Buła, P. (eds.) Management in the time of networks, cross-cultural activities and flexible organizations, pp. 147–162. Cracow University of Economics, Cracow (2017)
45. Lewandowski, R.A., Sułkowski, Ł.: New Public Management and Hybridity in Healthcare: The Solution or the Problem? In: Savignon AB, Gnan L, Hinna A, Monteduro F (eds) Hybridity in the Governance and Delivery of Public Services. Emerald Publishing Limited, pp. 141–166 (2018)
46. Lo Storto, C., Goncharuk, A.G.: Efficiency vs effectiveness: a benchmarking study on European healthcare systems. Econ. Sociol. **10**(3), 102–115 (2017)
47. Lounsbury, M.: Institutional transformation and status mobility: the professionalization of the field of finance. Acad. Manag. J. **45**(1), 255–266 (2002)
48. Mannion, R., Davies, H.T.: Payment for performance in health care. BMJ **336**(7639), 306–308 (2008)
49. Manzoor, Q.A.: Impact of employees' motivation on organizational effectiveness. Bus. Manage. Strategy **3**(1), 1–12 (2012)
50. May, C., Finch, T.: Implementing, embedding, and integrating practices: an outline of normalization process theory. Sociology **43**(3), 535–554 (2009). https://doi.org/10.1177/0038038509103208
51. McCormack, B.: Person-centredness in gerontological nursing: an overview of the literature. J. Clin. Nurs. **13**(s1), 31–38 (2004). https://doi.org/10.1111/j.1365-2702.2004.00924.x
52. McKay, K.A., Narasimhan, S.: Bridging the gap between doctors and nurses. JNEP **2**(4), (2012). https://doi.org/10.5430/jnep.v2n4p52
53. McPherson, C.M., Sauder, M.: Logics in action: managing institutional complexity in a drug court. Adm. Sci. Q. **58**(2), 165–196 (2013)
54. Meyer, J.W., Rowan, B.: Institutionalized organizations: formal structure as myth and ceremony. Am. J. Sociol. **83**(2), 340–363 (1977)
55. Mintzberg, H., Waters, J.A.: Of strategies, deliberate and emergent. Strateg. Manag. J. **6**(3), 257–272 (1985)
56. Moore, L., et al.: Barriers and facilitators to the implementation of person-centred care in different healthcare contexts. Scand. J. Caring Sci. **31**(4), 662–673 (2017). https://doi.org/10.1111/scs.12376
57. Nadeem, M., et al.: Impact of employee motivation on employee performance (A case study of Private firms: Multan District, Pakistan). Int. Lett. Soc. Humanistic Sci. **36**, 51–58 (2014)
58. Naldemirci, Ö., et al.: Deliberate and emergent strategies for implementing person-centred care: a qualitative interview study with researchers, professionals and patients. BMC Health Serv. Res. **17**(1):527 (2017). https://doi.org/10.1186/s12913-017-2470-2
59. Patient Safety: Danish society for patient safety. (2020). https://patientsikkerhed.dk/content/uploads/2016/08/dsfppersoncentredcare.pdf

60. Platis, C., et al.: Relation between job satisfaction and job performance in healthcare services. Procedia Soc. Behav. Sci. **175**(1), 480–487 (2015)
61. Ratanawongsa, N., et al.: What motivates physicians throughout their careers in medicine? Compr. Ther. **32**(4), 210–217 (2006)
62. Reay, T., Hinings, C.R.: Managing the rivalry of competing institutional logics. Organ. Stud. **30**(6), 629–652 (2009). https://doi.org/10.1177/0170840609104803
63. Reay, T., Hinings, C.R.: The Recomposition of an Organizational Field: Health Care in Alberta. Organ. Stud. **26**(3), 351–384 (2005). https://doi.org/10.1177/0170840605050872
64. Rigoli, F., Dussault, G.: The interface between health sector reform and human resources in health. Hum. Resour. Health **1**(1), 9 (2003)
65. Samuel, S., et al.: Monetized medicine: from the physical to the fiscal. Acc. Organ. Soc. **30**(3), 249–278 (2005)
66. Santana, MJ., et al.: How to practice person-centred care: a conceptual framework. Health Expect. **00**, 1–12 (2017). https://doi.org/10.1111/hex.12640
67. Scott, W.R., et al.: Institutional Change and Healthcare Organizations: From Professional Dominance to Managed Care. University of Chicago Press, Chicago (2000)
68. Shahzadi, I., et al.: Impact of employee motivation on employee performance. Eur. J. Bus. Manage. **6**(23), 159–166 (2014)
69. Silva-Ayçaguer, et al.: The null hypothesis significance test in health sciences research (1995–2006): statistical analysis and interpretation. BMC Med. Res. Methodol. **10**(1), 44 (2010)
70. Smaldone, P., Vainieri, M.: Motivating health professionals through control mechanisms: a review of empirical evidence. J. Hosp. Admin. **5**(3), 75 (2016)
71. Suddaby, R., Greenwood, R.: Rhetorical strategies of legitimacy. Adm. Sci. Q. **50**(1), 35–67 (2005)
72. Svensson, R.: The interplay between doctors and nurses—a negotiated order perspective. Sociol. Health Illn. **18**(3), 379–398 (1996)
73. Swarna Nantha, Y.: Intrinsic motivation: the case for healthcare systems in Malaysia and globally. Hum. Resour. Dev. Int. **20**(1), 68–78 (2017)
74. Thornton, P.H., Ocasio, W.: Institutional logics and the historical contingency of power in organizations: Executive succession in the higher education publishing industry, 1958–1990 1. Am. J. Sociol. **105**(3), 801–843 (1999)
75. Thornton, P.H., Ocasio, W.: Institutional logics. In: Greenwood, R., Oliver, C., Suddaby, R., Sahlin-Andersson, K. (eds.) The Sage Handbook of Organizational Institutionalism. pp. 99–128 (2008)
76. Thornton, P.H., Ocasio, W., Lounsbury, M.: The Institutional Logics Perspective: A New Approach to Culture, Structure, and Process. Oxford University Press on Demand (2012)
77. Valentjn, P., et al.: Understanding integrated care: a comprehensive conceptual framework based on the integrative functions of primary care. Int. J. Integr. Care (2013). URN:NBN:NL:UI:10–1–114415

Part III
Intelligent Systems and Their Application in Healthcare

Chapter 9
Overview of the Artificial Intelligence Methods and Analysis of Their Application Potential

Dalia Kriksciuniene and Virgilijus Sakalauskas

Abstract The medical industry collects a huge amount of data, most of which is electronic health records. These data cannot be processed and analyzed using traditional statistical or data analysis methods because of the complexity and a volume of the data. So the knowledge discovery from raw clinical data is a big challenge for healthcare system. In this chapter we introduce the issue of data mining in healthcare, i.e. how to use the raw clinical data to ensure a systematic approach to health problems, highlight good practices, reveal inefficiencies, and improve healthcare efficiency. We identify the data sources used in healthcare, discuss its adequacy, interpretation, transformation and cleansing challenges. Also we consider the variety characteristics and specific capacities of methods, applied in the areas of data mining. Particular attention is paid to the diversity of Machine Learning and Artificial intelligence methods, analytical health data analysis models, its testing and evaluation capabilities.

Keywords Artificial intelligence · Data mining · Healthcare data · Machine learning algorithms

9.1 Introduction

It is common, that business entities understand the importance of historical data and the role of analytics for exploring activities of a company or organization. A wealth of data on a variety of statistical and data mining techniques empowers to predict the future of a business or an enterprise, plan marketing and production strategies. Unfortunately, the amount of data collected does not yet ensure quality and value of analytical information to be provided to its owners. The "use of qualified data mining methods" would allow to obtain hidden information from the data and help

D. Kriksciuniene · V. Sakalauskas (✉)
Vilnius University, Vilnius, Lithuania
e-mail: virgilijus.sakalauskas@knf.vu.lt

D. Kriksciuniene
e-mail: dalia.kriksciuniene@knf.vu.lt

D. Kriksciuniene and V. Sakalauskas (eds.), *Intelligent Systems for Sustainable Person-Centered Healthcare*, Intelligent Systems Reference Library 205,
https://doi.org/10.1007/978-3-030-79353-1_9

to make the right decisions and predict prospects when there were uncertainties in real situations [1, 2].

Data mining is strongly based on classical statistical principles and general specific analytical techniques. However, their tasks are different. Data mining focusses on the applications domain- oriented goals of the performed analyses, whereas the classical data analysis methods are better understood and revealed by the data base, its properties, and key relationships. Thus, even the historical data-driven "black box" methods such as neural networks, swarm intelligence, and other machine learning methods are considered to be useful data mining methods to help predict the behaviour of analysed processes, although they do not enable to determine the nature or causal relationships of individual variables in the form of rules. From the point of view of classical statistics, such methods are considered to be 'dirty' ([3], p. 8). However, due to the abundance of practical applications, all these 'dirty' methods have taken their rightful place in the ranking among the classical methods.

The classical statistical methods strive to determine data structure by a predetermined model, while the data-driven models seek to discover structure from the acquired data. The first approach conforms to the Aristotle's deductive search: first a model is chosen and evaluated how the explored situation fits the theoretical method. The data mining models are more in line with Plato's inductive relationship of determining path-truth by gradually improving our approach based on the available information [2].

The healthcare data does not follow the predetermined structural requirements or processes, which are common for the business entities. The healthcare data is captured for big variety of purposes, starting from investigating health parameters (which are assigned to the precision medicine), it originates in the form of handwriting or voice records of doctors, or register visits and prescriptions. The variety of data sources and analytical goals makes the field of healthcare information analytics fragmented and unspecified by means of application of methods and evaluating their performance. The chapter analyses the characteristics of the data sources of healthcare domain and their enhancement due to the PCC approach, the solutions for their analytics, the general data mining process and considerations for building analytical models for healthcare.

9.2   Healthcare Data Sources

In general, the healthcare systems encompass the relationships among the individuals aiming to investigate their health status by employing the competences of medical personnel. This relationship is maintained by building complex web of interrelationships, which ensure all types of services, infrastructural support, financing, production of pharmaceutical and other material provision, as well as governmental and state-wide maintenance and control. There is no standard approach for implementing systems which could be able to reflect this complexity, therefore the healthcare systems worldwide are built and modified combining their unique settings, available budgets, best practices.

The information flows are emanating in each interaction of the healthcare system, making it natural to increasing pervasive application of information technologies. However, the human factor is strong and to high extent irreplaceable in all types of the interactions of the healthcare system which is created for people and driven by people. This makes it impossible to build the healthcare information system similarly to management information system concept as a "mirror of real processes", designed to observe and monitor its performance and efficiency by exploring process and transaction data entirely in digital mode. The digitalization of healthcare information only partially captures and covers the areas, which enable data collection and processing for analysis, inference, forecasting, and expert insights.

Summarizing the discussion of previous chapters of the book, the broad four groups of data sources can be identified in the healthcare systems in general, where the group 4 correspond to the specific requirements for application of PCC:

Group 1:   The intentionally collected exact measurements of healthcare parameters, such as heart rate, weight, blood structure, and other similar numeric indicators, visual information of X-ray, telemedicine information exchange, which have the established system of their measurement and application for healthcare processes, such as diagnostics, treatment, surgery, rehabilitation.

Group 2:   Collecting health related data and factors, as well as the healthcare outcomes and expert knowledge for elaborating data driven models with the goal to assist experts in decision making and their professional work processes. These data are collected for processing and partially replacement or supplementing competences and expert knowledge of medical specialists.

Group 3:   The healthcare economics data, such as census data, tax, budget, healthcare service and human resource cost and other information which influence performance and efficiency of the healthcare system. The design, evaluation and monitoring of the efficiency measures enable comparative evaluation and improvement of the healthcare systems in general, and their cost containment.

Group 4:   Searching and identifying initially unknown factors, causalities and insights which could potentially determine the health status of person and its deviations. These data are collected from multiple sources, both structured and unstructured, and are processed by integrated intelligent analysis. In general healthcare systems this type of analysis has supporting role to the needs of health promotion, however, it becomes core need for applying the Person-centred care approach.

9.3   Data Mining Solutions in Healthcare

The level of application of the data mining, machine learning and AI methods in healthcare is varied: some of the methods are already widely applied, especially for the data of Group 1. Analysis of the healthcare system based of PCC approach reveals no only the conceptual change. The idea of person centeredness enables continuous involvement of person in the healthcare system by consistent and broad capturing information related to health status (without targeting specific diagnosis), also increases responsibility and awareness of the person by increasing his healthcare literacy, which leads to the necessity to capture the information which characterizes life style, as well as the emotional context, family and community involvement in person care, as well as resource availability, motivation, and other factors. The identification of important data, access to its sources and complexity of its processing is especially high for the data of Group 4, which determines current low, yet consistently increasing level of applying AI.

The variety of AI methodologies applied for healthcare data analytics, as discussed in the systematic review [4], discovered that 65% (244/378 publications) of the relevant publications focussed on analysis of structured clinical data, followed by unstructured imaging data (17%). This systematic review included publications o AI application in healthcare on the period 2000–2018. The ANN and deep learning networks were the major methods applied in the research works. The search of recent articles in Web of knowledge 2018–2021 revealed dramatic increase of AI application in healthcare, comparing to the results reported in Ben Israel et al. [4]. The search of Web of Knowledge during 2018–2021 for indexes: SCI-EXPANDED, SSCI, A&HCI, CPCI-S, CPCI-SSH, ESCI and topic (healthcare AND ("artificial intelligence")) gave 1213 results, where 441 were assigned to computer science and health informatics disciplines. However narrowing the search to PCC approach by topic (healthcare AND ("artificial intelligence") AND ("person centered")) gave only one result, where the recurrent neural networks (RNN) were applied for questionnaire data. The search of Web of Knowledge for topic: (healthcare AND ("person centered")) during 2018–2021 gave result of 247 publications, however only 12 of them fell in the categories computer science artificial intelligence or computer science cybernetics or computer science interdisciplinary applications or engineering biomedical or computer science theory methods or medical informatics or information science library science. It can be summarized that application of AI methods for the data sources corresponding to the PCC approach is in the initial stage of research.

The variety of machine learning and AI methods can be revealed by applying several principles: classification by their paradigms, such as supervised, unsupervised, reinforcement learning, deep learning, and ensemble methods. The more widely accepted approach is based on summarized representation of AI algorithms (methods), and their modifications. In Fig. 9.1 the summary of machine learning and AI methods reveals their symbolic mapping, variety and constant new developments of this area. The new developments have brought new titles of algorithms

Fig. 9.1 The machine learning algorithms (https://machinelearningmastery.com/how-to-implement-a-machine-learning-algorithm/)

and new achievements in analytics. To name a few, the methods include Long short-term memory (LSTM), deep belief network (DBN), RPA (Robotic process automation), (https://www.guru99.com/robotic-process-automation-tutorial.html), and its enhancement in the AI area to Intelligent Process Automation (IPA), language processing GPT-3.

We can consider the variety, characteristics and specific capacities of methods, generally applied in the areas of data mining, processing, analytics and computational intelligence, mainly defined as the areas of machine learning and Artificial intelligence (AI):

- Classical statistical research methods (Regression, ANOVA, discriminant analysis);
- Logistic and probabilistic regression; Classifiers and statistical learning methods.
- Supervised and unsupervised learning methods (Neural networks and clustering)
- Association rules
- Decision trees
- Fuzzy logic
- Genetic algorithms and swarm intelligence.

Clearly, the application of these methods is not limited to healthcare tasks. They are very widely used in economics, finance, meteorology, marketing, industrial process management, and other domains [3, 5–7]. There are numerous cases, where the AI methods have broader application in the domain areas, which deal with the non-person related data. The industrial data has of lower sensitivity level and ethical risks for its processing, therefore in many cases it becomes pioneer domain for testing new AI methods. Successful application of the methods in various areas, as well as the tendency of rapid development of the modified and integrated solutions makes it possible to transfer the research to the medical domain. The recent achievements and solutions for cybersecurity, ethics, anonymity and others assist to more rapid AI application process in the healthcare domain.

In the following part we will present a general concept of the methods and illustrate their application in medical informatics, healthcare and person centred problem areas.

9.4   Development of Data Mining Process

Data mining in its broad sense refers to approaches and methods used for collecting and analysing sufficiently large amounts of testing data, allowing us to identify the significant relationships, trends and rules hidden in the data, and to apply the discovered regularities to new subsets of data with the same characteristics. The latter task—anticipating future prospects—is one of the most important tasks in data mining and is highly needed for application in various areas of healthcare data processing.

Application of data mining for solving healthcare tasks could enable doctors select best treatment plans, ensure the quickest healing process, increase patient satisfaction, or acquire better knowledge of their customers for its future application in healthcare processes. Many trends in patient behaviour or the development of health conditions only become apparent over time. Do the patients come back often, what health related habits they have to take into account, what offers and advice they are most interested in?

These are natural questions that any every healthcare service provider should know the answer to. In order to provide reliable answers to similar questions it is necessary to collect data about the patients, their health parameters, symptoms, features of diseases to be analysed for diagnostic, treatment and other purposes. The appropriate data mining methods have to be applied to analyse all relevant information. The essential assumption of data mining discipline is that past data encompasses useful information for the future. Data mining will help us to separate the meaningful signal from the noise, find the relationships between the variables, determine the trend characteristics of their change, empower with the inferential knowledge, assist decision making and forecasting.

Many data mining methods are designed for model building. A model is understood as an algorithm or set of rules that combines input variables with the output or target variables. The model should explain how a particular outcome may be determined by certain observable circumstances (input variables) and predict the outcomes in the same or in an analogous situation [8, 5]. For example, a model for predicting the outcome of the person illness the doctor should select and acquire for analysis a lot of the most relevant input variables, characterizing important symptoms and identifying diagnosis of the sick person, such as physical, chemical and structural

Fig. 9.2 Application of data
mining model development
process

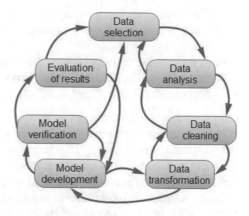

characteristics of blood, muscles, brain, presence of other illnesses, life style habits, emotional status, etc.) and define their relationship to the target variables, such as health recovery level after treatment.

The process of model development by using data mining techniques generally combines the following steps [9]:

1. Select data suitable for the analysis of the problems;
2. Perform initial data analysis;
3. Cleanse the data;
4. Transform the data;
5. Create a model;
6. Check the model;
7. Get results.

The steps of model development can be performed in any order. They can be used interchanging, and even repeated. The insights and knowledge which is discovered in later steps may require to repeat and take corrective actions in the previous ones. The Fig. 9.2 presents data mining model development process as a cycle containing of the tightly interconnected steps.

In the following sections the data mining model development process steps will be analysed in more details.

9.5   Data Sources

Data availability, its sufficiency and quality is essential for building analytical models of high reliability and precision. Data may be collected on-purpose and stored in the information systems for tracking financial operations, transactions, provision of services, or getting survey results on specific questions. Selecting, acquiring and managing data is a complex task:

- The amount of data is growing exponentially, old data needs to be stored, new data are constantly emerging, making it difficult to find the data needed for each solution
- Data records are scattered across different sources, stored by different methods and devices, and stored on different servers or processing systems.
- Different formats, encodings and presentation options are used.

Data from internal data sources is usually stored in the company's own databases, either in-house or cloud environments. Their content is addressing existing or potential customers, products, services and processes. The employees generate lots of information, which is in many cases not consistently documented or stored by the institution. It concerns competence and professional knowledge, concepts, thoughts, opinions, subjective evaluations of products and services, insights about competitors) which is often collected and maintained by the employees themselves. Part of the information related to operations and work practices, such as user-created rules, formulas, models are stored in the company's knowledge base.

The data sources, which cannot be managed or intentionally collected by the company for its specific purposes are generally referred as external sources. However, they may contain useful and influential data affecting the explored indicators. The examples of the external sources are databases of healthcare statistical information, civil registries, databases of legal documents, as well as the information coming in visual, audio, comment or recommendation forms form social networks, surveillance cameras and others. The monitoring and analysis of the external data sources is necessary in the stage of data selection for analytical purposes, as the available data may be not related or only partially related to the pursued objectives and problem areas.

The data can be captured and registered in the databases automatically or by the input of users, customers, or person respondents. In healthcare the data input automation can employ devices such as ambience or body sensors, observation cameras, magnetic resonance scanning and computerized tomography diagnostics devices. The manual input is widely applied as well due to the requirements for doctors and nurses to register observation and interview information with the patients, checking their personal information during admission, registering outcomes of the patient visit, including recommendations for treatment, pharmaceutical and rehabilitation purposes. The healthcare or governmental institutions may initiate collection of survey data, expert knowledge, observation materials.

Broad variety of technical and computational tools may be applied for capturing various kinds of data, its transformation and transfer for analytical purposes. The common situation of necessity to combine data sources of different formats and origins adds complexity to the data selection procedure, where specific environments, especially providing services of Big data scale, can be employed for preparing data in the forms which are ready for further processing.

The data selection process can imply technical difficulties for data acquisition. It also affects the data quality and the reliability of further analysis. The main drawbacks of data sets maybe summarized as a problem of "dirty data", which means that the information may be not fully trusted, it may be not full, or can contain duplications and contradictions while interpreted in different contexts. The decision to apply low quality data for the data mining process bring loss [10].

The data management extent in the enterprises, healthcare, public, governmental institutions is pervasive and constantly growing. It has brought many regulations and requirements for data owners for ensuring data reliability, safety, integrity, recovery, accessibility, ethical, cybersecurity and others. Different domain areas and countries may apply specific and different legal requirements for this purpose. The medical data experiences influence of these kind of regulations as well, stating from protecting sensitive personal data, defining its sharing among healthcare institutions or providing rights of their editing, access via the user identification and authorization processes.

9.6   Tasks of Data Analysis

In order to select the data suitable for the analysis of the problem, we should first take into account the origin of the task we are solving. In general, the tasks may tackle:

- Classification;
- Evaluation;
- Prediction;
- Affinity grouping and association rules;
- Clustering.

Classification is understood as the allocation of objects into the predefined classes. The classification task needs a precise explanation of the classes, and a training set covering pre-classified instances [5]. Classification operates with a discrete result, nominating or coding the identified classes. In healthcare, the diagnosis classification codes are used as a tool to group and identify diseases (such as cardiovascular, diabetes mellitus, flu), and other reasons for patient encounters, such as injuries or adverse effects of drugs. which enables to assign the most credible defined diagnosis into a code from a particular classificator. Decision tree, Neural networks, Nearest neighbour, Support Vector Machines are the main techniques and algorithms often used to solve classification problem [5].

Evaluation problem is built by selecting several input variables for estimating the solution of the problem, thus leading to the continuous numeric result, which generally falls into the interval [0,1]. This type of problem can be solved for estimating rating of patient 0 and 1, where 1 indicates a completely reliable parameter value, and 0 indicates the completely unreliable one. The physical fitness estimation is provided with the goal of obtaining overall health screening, including the level of cardiorespiratory form, the level of muscular system strength and endurance, or the level of flexibility. Regression analysis and neural networks can be used to solve evaluation tasks. This approach can also be used to estimate the time to an event (loss of a patient), where the survival analysis is applied.

Prediction is used to predict the future value of an output variable. This can be used for both classification and evaluation of the explored variable, only the time feature is added for relating available data in temporal sequence. We can check the accuracy of the forecast only after reaching the forecasted future and comparing the difference between the forecast and the actual value. Any classification and estimation techniques can also be used for forecasting by taking training data that contains the known values of the predicted variable. We can predict and forecast the number of patients to be admitted to the hospital during the coming month, the amount of the material resources to be ordered for use in the intensive care departments, the number of people who will get sick with flu during autumn season.

Affinity grouping denotes tasks where we aim to find objects or events that usually appear together. This grouping method is commonly used in diagnostics or observations for deriving insights and rules: it could evaluate if a particular coronary disease goes together with smoking, if the sleeping duration of less than 6 h daily and short-term stress is observed together with occupational burnout syndrome, does the majority of people buy prescribed medications together with self-selected vitamins. The method of association rules is often applied for affinity grouping tasks.

Clustering, unlike classification, segments objects into the previously unknown classes. Merging into clusters is performed based on the similarity of the objects to be merged. Similarity is determined by taking into account all selected characteristics of the object registered as a set of variables. Only a specialist conducting the study can determine the meaning of the resulting cluster according to the context emanating from the available input variables. It may include tasks of clustering people according to the different lifestyle habits (e.g. smoking, low physical activity) or their emotional characteristics to different segments and lead to the application of different rehabilitation tactics after their illness. Clustering by various complaints, ailments and pains can lead to recognizing segments of citizens with particular risk factors or excessive healthcare cost usage. The technique used for clustering is based on the estimating similarity of the grouped objects according to the selected metrics. Self-organizing maps method have recently become especially popular in solving clustering tasks.

The first three tasks of Classification, Evaluation and Prediction (Forecasting) make a category of the directed (supervised) data mining tasks. They always contain the goal variable (or known output) that we aim to classify, evaluate, or predict. Affinity grouping and clustering make a category of the undirected (unsupervised) data mining tasks [5]. In these cases, the input data set has no goal variable. Our objective is to find general regularities that are not directly related to a specific variable (Fig. 9.3).

Once the research task has been set, it is necessary to decide what data would be needed to obtain significant results. It is beneficial to have a larger number of variables at the beginning of the data mining process, and only then refine research

Fig. 9.3 Classification of
data mining tasks

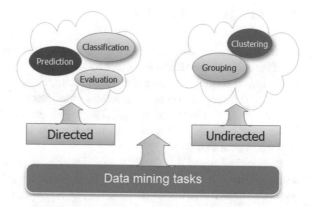

by selecting the most needed ones. What data are important and how much of it is needed can be decided by the expert researcher, taking into account the importance of the problem, the possibilities and cost of data collection, the subject area studied and the required accuracy of the final result.

Generally, the efforts are directed to collect as much data as possible. Only in this way can sufficient accuracy of the developed model be ensured. When there is too little data, the outcomes of the research can be ineffective or perhaps even completely pointless. If the efficiency of the model developed with a larger sample significantly bigger than using a smaller sample, it means that a smaller sample is certainly insufficient for the elaborated model. If the difference between the efficiency levels is insignificant, we can use a smaller amount of data for applying the model. The problem of data set size is important, a study using a large sample may require significantly more computing time and costs for data acquisition.

The selected prediction or forecasting technique of data mining aims to characterize future by past data, therefore it is important to have data for quite a long past period. Most of the past data has similar behaviour related to specific days of the week, weekend and holiday effects of seasons of the year, which are described as seasonality component. In general, it is recommended to have data for at least 4–6 historical seasons, but in specific cases more adjusted data requirements may be applied.

For some data mining tasks, the number of variables available for the research can be quite low, and it can negatively influence the performance of the data driven data mining methods. In these cases, the derived variables can be created, which present the summarized characteristics of data, adding value for the research. From the patient record database consisting of the information on each patient visit, we can calculate the frequency of visits, the age at particular occurrence of illnesses, the deviation from recommended body mass index at particular age, or average number of days at hospital. Thus adding new features for analysis, and providing overall picture of the aspects included to the research problem area can help better understanding the research problem.

9.7   Adequacy of the Data

Before creating a model, it is very important to perform data analysis to determine the adequacy of the data we collect for the task at hand. First of all, it is helpful to explore different visualisations of the data, draw a histogram of each variable, calculate the most important numerical characteristics, enabling to reflect trends, anomalies, or potential groupings. The initial exploring of data allows to detect incorrect data coding, illogical data instances, or inadequate data spread along the explored variable range.

It should be noted that the data would maintain the distributions and proportions within existing study area. If the survey is conducting for reflecting opinion of the region, it should be ensured that the survey covers a relevant proportion of respondents by age groups, gender, the ratio of urban to rural population, or the percentage of married people does not differ much from the statistical characteristic of the region.

It is always necessary to check whether the specific values of the variables meet the prescribed limits or whether the interdependence of the variables corresponds to reality. Given that the unemployment rate of some the country is 9%, a similar number of unemployed should be expected among our respondents.

The in-depth knowledge of data may be helpful and important in the cases of values, which may change their meaning over time. In the case of ratings, the rules for assigning the highest category may have experienced modifications, therefore the same rating "A" may mean different thing in different time periods [11]. In 1998, the U.S. National Institutes of Health and the Centres for Disease Control and Prevention brought U.S. definitions in line with World Health Organization guidelines, lowering the normal/overweight cut-off from BMI (Body mass index) 27.8 to BMI 25. The decision reclassified 29 million U.S. person, previously healthy, to overweight [12].

Data Cleansing

At this stage, the data need to be carefully reviewed for any coding errors, missing values or formatting mistakes in the data set, that could severely hamper the application of a particular data mining method need to be corrected. Neural networks are particularly sensitive to data errors. Decision trees and classification techniques are less sensitive to data cleanliness [13].

A common need in this stage is to decide about the data records that have non-existent, missing values. Several approaches can be applied: firstly, the records with the missing values may be not included, but then the data file may become insufficient, the deviation by proportion and number of records from real situation can occur. Replacing non-existent values with a similar mean-median or median has risk of distorting the real situation, if the existing values do not reflect the mean value of the collected data set. It is proposed to take into account the applied data mining method and the problem to be solved when choosing the method of analysis of non-existent values.

Serious problems arise when the target variable is categorical, and may acquire many values. Such types of variables are the education levels, possible cities of residence, zip code, and similar multi-valued characteristics. In this case, classification or prediction is quite difficult because some of the acquired values may lack information to describe them due to lack of records with this data value. An attempt should be made to pre-group the values of such variables or replace them with values of interest to us. In this way, the diversity of all education levels could be changed into primary, secondary and high.

Sometimes there are variables with several values that are very different from all the others. This difference can reach tens or hundreds of times. The inclusion of such values in the study can severely skew the result. Analysing the resources, available for treatment of the retired patient the statistics of Lithuania indicates differences among the minimal and maximal monthly pension approximately 9,09 times, and among the EU citizens the difference in even more crucial: as of 2018, Luxembourg recorded by far the highest level of annual median equalised net income among older people (aged 65 years or more), at 35 101 PPS (purchasing power standard), making the difference to lowest range countries, namely Lithuania, Latvia, Romania and Bulgaria, average income for older people (5749 PPS) (Eurostat, 2018).When examining such variables, it is advisable to transform such a variable by replacing each of its values with a logarithm, or even to abandon the exclusive records, if the records were occasional, and their inclusion is not necessary for the problem, e.g. proportional representation of the explored countries.

Data Transformation

Data cleansing is followed by the step of data preparing for analysis. Many data mining methods do not have the ability to work with particular data fields. Thus, they may need to be transformed into an understandable format or replaced by a variable of duration. Sometimes it is enough to replace a temporal variable with a categorical variable. Let's say we have information about the date of birth of a school child. In many types of problem, such as evaluating physical activeness (PS) the most informative transformation may be sufficient to use a derived categorical variable to categorize patient by age: Preschool-Aged Children (3–5 years), Children and Adolescents (6–17 years), instead.

It is advisable to seriously consider what additional variables may be needed before starting model development. Various derived indices and percentage comparison of variables are very useful in research. Here are some examples:

- Body mass index = Body weight (kg)/(Height (m))^2
- Life Expectancy Index (GTI) = (Life Expectancy-25)/60
- PE – Price/Profit

It is not recommended to use the variables expressed by the frequency. It is better to replace them with a relative frequency or a percentage.

9.8 Building Analytical Models

This is the most important step in data mining. Its application depends on the specific method we use. In the following sections, describing the individual methods, we will present the essential characteristics and application of the modelling algorithms. In the cases of directed data mining tasks, the training set is used to explain the behaviour of the outcome variable depending on the input variables. This type of interpretation may be done applying neural networks, regression analysis, genetic algorithms, decision trees, or other methods capable to model the outputs by the input data set.

While solving the undirected data mining problems, the developed model must determine the relationships and dependencies between the input variables and express them through the dependency rules or by clustering the variables [14].

Data mining techniques are often classified according to the nature of the method used. There is a distinction between the supervised and unsupervised learning modes. The supervised learning is understood as a method which explores relationships and dependencies among data arranged as variety of input variables and the known output parameters. The unsupervised learning methods enable to draw conclusions only from the training data without knowing the output variables.

Examples of supervised learning are: Classification tasks; Applications of regression analysis; Time series forecasting; Various optimization tasks.

Unsupervised learning methods could include: Factor and principal components analysis; Clustering tasks; Interoperability rules.

The new models and the improvements of existing algorithms is constantly emerging in both classes.

9.9 Model Testing

In order to check whether the created model properly describes the observed situation, we should know, as precisely as possible, what we are aiming for, what precision and accuracy of the results do we need, what is the sensitivity and flexibility of the model and the possibilities to adapt to the changed conditions. It is clear that the answer to these questions depends on the type of model developed. Let's say models related to human disease or high costs require increased reliability.

The efficiency of the directed data mining models is tested by using a set of test data which was not used for model development. In the classification and forecasting tasks, the accuracy of the model is measured by the error rate-percentage of incorrectly classified or predicted records. In the evaluation tasks, accuracy is understood as the difference between predicted and observed results. The overall accuracy of such models is measured by the mean error of the individual estimates. Since the average between the predicted and observed values is always equal to 0 (deviations of the sign to be distinguished outweigh each other), the average of the differences from the square deviations is usually taken [15].

The standard error of the estimate can be calculated by the formula:

$$MSE = \frac{1}{n} \sum_{i=1}^{n} (r_i - p_i)^2$$

where MSE denotes Mean Squared Error, r_i—the observed value, p_i—forecasted or predicted value, and n—number of observations. The lower value of MSE means more accurate results provided by the mode.

MSE has several main drawbacks. Firstly, this measure has high sensitivity to presence of even small number of values with big deviations. It can happen that even one anomalous values of the data set may result to misleading final results. The drawback of MSE can be solved by applying MAE-Mean Absolut Error:

$$MAE = \sum_{i=1}^{n} \frac{|r_i - p_i|}{n}$$

However, the MAE evaluation of error is not often applied due to computational inconvenience. Another drawback of the MAE is the higher rate of the absolute value of error, which can be eliminated by square root of the MSE. In this way we get the error evaluation method SD-Standard Deviation:

$$SD = \sqrt{\frac{1}{n} \sum_{i=1}^{n} (r_i - p_i)^2}$$

It is more difficult to evaluate the precision of the undirected methods. In these tasks the most important outcome is its power to provide relevant description of various characteristics of the explored problem. The most optimal model can be understood as a sufficiently small number of rules, able to provide most complete explanation of the behaviour of the model. The measure which can express efficiency of the model is MDL-minimum description length and is defined as number of bits necessary to code all rules defining the mode. This measure can be used to compare several sets of rules among themselves, and to select the best one according to smaller MDL.

9.10   Evaluation of the Data Analysis Results

As the data mining processes are highly sensitive to the quality and amount of the analysed data, it may happen that the developed model is able to accurately describe the future situation, but its practical implementation could take unfeasible amount of time or require inadequate costs. In this case, it can be considered whether it is worthwhile for us to try to implement this model in practice. Value of the solution, its costs and time are the indicators that determine the expediency of applying the model and the possibility of practical application of the results. Particular care should be taken in evaluating the results for using diagnostic, treatment, rehabilitation resources, including human resource, as well as pharmaceutical medication selection, as all such costs are included in the total cost of the healthcare institution.

Unfortunately, the reviewed steps in the data mining study do not ensure the satisfactory investigation process. It is highly probable that the developed model raises more questions than it provides answers. That could indicate a valuable situation, that the data mining efforts have helped to uncover new previously unforeseen relationships, or discover particularly sensitive characteristics of the available data. In this case the improvements of the model can be achieved by acquiring new set of data, modifying, integrating or choosing new methods for analysis. Consequently, the new level and breakthrough can be achieved for preparing advanced solutions to the problem.

9.11 Conclusion

The data mining and artificial intelligence methods demonstrate vast potential for analysis in the areas of generating large amount of data, both structured and unstructured. However, the problems of pattern recognition, knowledge discovery from healthcare data are still in their initial stage, mostly oriented to structured measurement data, and narrow number of data mining methods. There are no defined procedures which methods are best fit for the analytical tasks. Introduction of basic principles of data mining and main artificial intelligence techniques, discussion of building, verifying and testing the data mining models will allow healthcare professionals to realise the importance and significance of Data Mining, Machine Learning and Artificial Intelligence methods for knowledge extraction from health data.

Acknowledgements This publication is based upon work from COST Action "**European Network for cost containment and improved quality of health care-CostCares**" (CA15222), supported by COST (European Cooperation in Science and Technology)

COST (European Cooperation in Science and Technology) is a funding agency for research and innovation networks. Our Actions help connect research initiatives across Europe and enable scientists to grow their ideas by sharing them with their peers. This boosts their research, career and innovation.

https://www.cost.eu

References

1. Larose, D.T.: Data Mining Methods and Models, IEEE Computer Society Press, p. 344, (2006). ISBN-13 978–0–471–66656–1
2. Nisbet, R., Elder, J., Miner, G.: Statistical Analysis & Data Mining Applications. Elsevier, Canada (2009)
3. Pregibon, D.: Data Mining. Statistical Computing and Graphics **7**, 8 (1997)
4. David Ben-Israel, W., Jacobs, B., Casha, S., Lang, S., Won, H.A., Ryu, M., de Lotbiniere-Bassett, D.W.: The impact of machine learning on patient care: a systematic review. Artif. Intell. Med. **103**(101785), 2020. ISSN 0933–3657. https://doi.org/10.1016/j.artmed.2019.101785
5. Berry, M.J.A., Linoff, G.S.: Mastering Data Mining. Wiley, New York (2004)
6. Edelstein, H., A.: Introduction to data mining and knowledge discovery, 3rd edn. Two Crows Corp, Potomac, MD (1999)
7. Han, J., Kamber, M.: Data Mining: Concepts and Techniques. Morgan-Kaufman, New York (2000)
8. Adamo, J.M.: Data Mining for Association Rules and Sequential Patterns. Springer-Verlag, New York Inc (2001)
9. Dunham, M.H.: Data Mining: Introductory and Advanced Topics, p. 315. Prentice-Hall, Pearson Education Inc. (2003)
10. Engelbrecht, A.P.: Computational Intelligence an Introduction, University of Pretoria South Africa, 288 p. John Wiley & Sons, Ltd (2002). ISBN 0–470–84870–7
11. Pyle, D.: Data Preparation for Data Mining. (1999)
12. CNN: Who's fat? New definition adopted. CNN. (1998). www.cnn.com/HEALTH/weight.guidelines. Archived from the original on November 22, 2010. Retrieved 2010–04–26
13. Verikas, A., Gelžinis, A.: Neuroniniai tinklai ir neuroniniai skaičiavimai.(In Lithuanian language) – Kaunas: Technologija, 2008. 241 p. ISBN 978–9955–591–53–5
14. Negnevitsky, M.: Artificial intelligence: Aguide to intelligent systems. Harlow, England: Pearson (2002)
15. Krikščiūnienė, D., Sakalauskas, V.: Intelektiniai modeliai marketingo sistemose Monograph. Vilnius, Vilniaus universiteto leidykla, 384 p. (2014)
16. Incomes for older people. Eurostat https://ec.europa.eu/eurostat/statistics-explained/index.php?title=Ageing_Europe_-_statistics_on_pensions,_income_and_expenditure. Accessed 10 Jan 2021

Chapter 10
Discovering Healthcare Data Patterns by Artificial Intelligence Methods

Dalia Kriksciuniene, Virgilijus Sakalauskas, Ivana Ognjanović, and Ramo Šendelj

Abstract The variety of the artificial intelligence and machine learning methods are applied for data analysis in various areas, including the data-rich healthcare domain. However, aiming to improve health care efficiency and use the captured information to improve treatment methods is often hampered by poor quality of medical data collections, as high percent of health data are unstructured and preserved in different systems and formats. In addition, it is not always agreed which methods of artificial intelligence and machine learning perform better in different problem areas, and which computer tools could make their application more convenient and flexible. The chapter provides essential characteristics of methods, traditionally applied in statistics, such as regression analysis, as well as their advanced modifications of logit, probit models, K-means, and Neural networks. The performance of the methods, their analytical power and relevance to the healthcare application domain is illustrated by brief experimental computations for investigation of stroke patient database with the help of several readily available software tools, such as MS Excel, Statistica, Matlab, Google BigQuery ML.

Keywords Artificial intelligence · Data mining · Healthcare data · Machine learning algorithms

D. Kriksciuniene · V. Sakalauskas (✉)
Vilnius University, Vilnius, Lithuania
e-mail: virgilijus.sakalauskas@knf.vu.lt

D. Kriksciuniene
e-mail: dalia.kriksciuniene@knf.vu.lt

I. Ognjanović · R. Šendelj
University of Donja Gorica, Podgorica, Montenegro

10.1 Introduction

In the age of computer technology, there is no shortage of data for analysis. The main problem is to decide what research method to apply, what insights we can get from this data and what decisions they can propose.

Typically, the analysis of a data begins with the application of classical methods of descriptive statistics and visualisation of data, which can help discover a data pattern or show trends in data change. One of the initial data analysis steps is to determine measures of central tendency (mean, median, mode) or measures of variability (standard deviation, data width, variance, asymmetry factor, excess). These characteristics provide a better understanding of the nature of research object and provide an initial picture of the data, their layout, quality and completeness.

The characteristics of data can be easily discovered by using a variety of computer programs, starting from the widely accepted MS EXCEL to specialized statistical calculation environments such as SPSS, STATISTICA, Matlab or Google BigQuery. For solving more advanced research problems, we will use different software and computer tools that will allow the reader to consider the most appropriate solution for a specific artificial intelligence problem.

The discussion and comparative evaluation the artificial intelligence approaches, and the illustration of their performance by applying different AI methods and tools should help us to reveal advantages of artificial intelligence and machine learning methods in the area of application of health data analysis in different cross-sections. This research topic is very popular and attract attention of many researchers [1–7]. For this purpose, we will take a big real clinical record file and try to analyse it using various research methods.

The database applied for the experimental research is a collection of registered stroke cases of the neurology department of Clinical Centre in Montenegro. The database consists of the structured records of 944 different patients, 58 variables, where 50 of them are coded by scale values of [1, 8, 9] corresponding to "Yes, No, Unspecified" conditions, and 8 variables consisting of the demographic data, admission date and discharge date from hospital. The data is collected between 02/25/2017 and 12/18/2019. The demographic data of stroke patients varies by age (from 13 to 96 years), and gender (485-male, 427-female).

Further, we will introduce several data research methods letting us to examine the structure of the data, find important patterns and disclose the relationships of the most important variables. We will try not only to present various research methods, but also will explain how to clean and transform the original data according to the task requirements, and to use different software tools for specific artificial intelligence and machine learning methods.

The next section will focus on understanding regression and correlation analysis and analysing the dependence strength of our data.

The Sect. 10.3 will examine logit and probit regression application to predict the variable *Vital_Status* of the stroke patient from different individual characteristics, such as *Type of stroke, Treatment methods, Health modified ranking score before stroke, Age at stroke* and *Gender*. Here we will introduce Google BigQuery Machine Learning capabilities to address this type of challenge.

The Sect. 10.4 describes the unsupervised machine learning method k-Means. This method let us partition data records to the predefined number of clusters. The calculations will be performed by the help of Matlab software.

The Sect. 10.5 explored application of neural networks for the supervised learning case of classification, by applying STATISTICA Data Mining tools.

10.2   Correlation and Regression Analysis

In general, regression analysis is a statistical method that allows the estimation of dependence among two or more quantitative variables in order to predict a dependent variable [10].

The simplest regression dependence is linear: $y = \beta_0 + \beta_1 x$. The coefficients of equation are found by the least squares method, i.e. minimizing differences between the points (x_i, y_i) and the regression curve. The regression analysis methods are very widely used in medical research. Usually, to draw regression line and calculate determination or correlation coefficients we can use MS Exccl software, but here we will apply the STATISTICA software package and limit our analysis to providing an example of the simplest regression curve.

We will illustrate the task of finding interdependence among the number of days spent in hospital and the age of the patients at stroke, which varied between 20 and 50 years, in Table 10.1 there is the sample of data set used.

Firstly, we explore a scatterplot for visualisation of data and finding a linear regression equation (Fig. 10.1).

As from Fig. 10.1, the linear regression equation for variables denoting age and days at hospital is: $Days\ At\ Hosp = -4,9906 + 0.4102 \cdot Age$. It enables to estimate forecast number of days to be spent at hospital according to the age at stroke. The relevance of the results, and its suitability for forecasting is judged by the coefficient

Table 10.1 Example of the data records

Days at hospital	Age	Days at hospital	Age	Days at hospital	Age
16	50	3	48	3	48
13	49	24	49	24	49
23	49	34	49	34	49
0	48	19	47	19	47
3	50	1	48	1	48
6	49	8	47	8	47
11	50	0	48	0	48
6	50	47	48	47	48

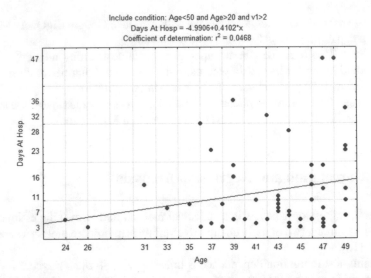

Fig. 10.1 Scatterplot for visualizing linear regression between *Age* and *Days at Hospital*

of determination. In our case (Fig. 10.1) the determination is equal to $r^2 = 0.0468$. The coefficient of determination is interpreted as the proportion of the variance in the dependent variable that is predictable from the independent variable in the range from 0 (no dependence between variables) to 1 (indication of a perfect fit). In the solved example only approximately 5% of variation of the dependent variable *Days At Hosp* can be explained by using the independent variable Age.

If the relationship between the variables is not well-fitted to linear (as in Fig. 10.1) we may use the non-linear regression. Then, instead of a line, we explore parabola, exponential, or logarithmic equations, and determine their unknown parameters by the least squares method. If a dependent variable is not well predicted by a single variable, several independent variables can be used to more accurately describe the situation. This type of regression is called group regression. Typically, group linear regression uses no more than 5 or 6 additional variables. Both group and curve regression calculations can be performed using the software tools already mentioned.

10.3 Logit and Probit Models

Traditional regression methods sometimes have difficulty describing a dependent variable that acquires values only from the range [0,1] or values of 0/1 (true/false, success/failure, error/non-error, etc.). In this case, logit or probit regression [11] are appropriated. The main difference among these two models is the different link function. The logit model uses cumulative distribution function of the logistic distribution, and probit invoke the cumulative distribution function of the standard normal distribution. Both functions may take any number as input, and rescale it to fall within the range of [0; 1]. These regression dependencies are applied for in medical, social science tasks, and are widely used to solve marketing and financial problems.

In order to illustrate the performance of these models we chose an example task, how the condition of blood pressure (0-normal, 1-high) may depend on age, weight, physical activity and stress level of the patient. Other similar examples could be evaluation of prostate enlargement (0-enlarged, 1- normal) from the available health indicators of the patient.

The basic assumptions of the logistic regression model are defined [12]: suppose the dependent variable y acquires a value of 1 with probability p, and it acquires a value of 0 with the probability of $q = 1 - p$. The types of independent variables for building logistic regression model can take any values, i.e. quantitative, qualitative, or categorical. The distribution of the input variables is not restricted for this model either.

In the logistic regression, the relationship between the outcome variable and the descriptive variables is not a linear function, as it was in the case of linear regression. The model of Logistic regression correlates probability value p with the independent variables $x_1, x_2, …, x_n$:

$$p = \frac{1}{1 + e^{-(a+b_1x_1+…+b_nx_n)}},$$

where a is a constant, b_i—the regression weights of the independent variables.

This equation takes another form after applying the logistic transformation function (logit) [12].

$$Logit(p) = \ln\left(\frac{p}{1-p}\right) = a + b_1x_1 + b_2x_2 + \cdots + b_nx_n.$$

The link function of the logit regression is expressed by $f(x) = \frac{1}{1+e^{-x}}$, and link function of probit regression is a function of the standard normal distribution $\Phi(x) = \int_{-\infty}^{x} \frac{1}{\sqrt{2\pi}} \exp\left(-\frac{y^2}{2}\right) dy$, which only slightly differs from the logistic one (Fig. 10.2).

Thus, we can define probabilistic regression as follows:

$$p = \Phi(Z) = \int_{-\infty}^{Z} \frac{1}{\sqrt{2\pi}} \exp\left(-\frac{x^2}{2}\right) dx,$$

where $Z = a + b_1x_1 + b_2x_2 + … + b_nx_n$.

The logistic and probabilistic regressions differ only by the transformation function, which determine differences of the behaviour of these models. The normal distribution function grows faster than the logistic one, therefore it provides a higher sensitivity to probabilistic regression, i.e. dependence on descriptive variables [13].

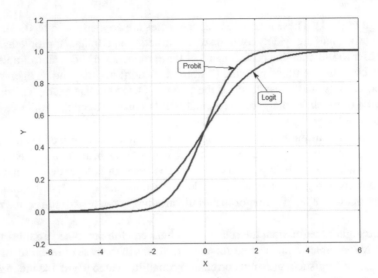

Fig. 10.2 Logit and probit link or transformation functions

The logit and probit regression models belong to the class of the supervised machine learning techniques. It means that the training set with the labelled examples is available for building the model. A supervised learning algorithm analyses the training data learning input/output regularities and produces an inference function, which can be used for estimating output for the new input examples.

We will solve the illustrative example of logit regression with help of Google BigQuery ML (see [8]). BigQuery ML enables to create and execute machine learning models by using standard SQL queries and the ML libraries. BigQuery ML supports not only the linear and logistic regression models, but also provides tools to apply K-means clustering, Matrix factorisation, Time series, Deep Neural Network and other computational intelligence methods.

The 944 data records of patients diagnosed with stroke were used for estimating logit and probit models. We will explore the "Vital status after hospitalisation" (1-Alive, 0-No) as a dependent variable, which is possibly affected by 5 independent variables: (1) Type of stroke, (2) Treatment methods, (3) Health modified ranking score before stroke, (4) Age at stroke, and (5) Gender.

In Table 10.2 the excerpt of data transactions and corresponding variables are presented.

Variable Stroke_Type gains value 1 for the diagnosis Ischemic stroke, 2 for Hemorag, 3 for SAH and 4 for unspecified stroke. The variable Vital_Status after hospitalisation may take values of Alive marked by 1, or not alive- 0. Variable Treatment_Methods denote categories of medications, or their combinations, received during the hospital stay, the corresponding values are in Table 10.3.

For example, the code value 13 means combining two types of medication Anticoagulation and Thrombolysis, 24-Dual Antiplatelet Therapy and medications from the broad group Other. Health_Status is evaluated from 0 to good health to 6-very bad

Table 10.2 Example of the data

Vital_status	Stroke_type	Treatment_methods	Health_staus	Age	Gender	Data_frame
1	1	24	0	60	1	T
1	1	2	0	59	2	P
1	1	24	0	59	1	P
1	1	24	0	58	1	P
1	1	24	0	58	2	P
1	3	0	1	58	2	T
0	3	4	0	60	1	T
1	1	24	0	58	1	T
0	1	14	0	59	2	T
1	1	2	0	58	1	T
0	1	24	0	57	1	E
0	1	24	0	57	1	P
1	1	24	0	57	9	T
0	1	14	2	58	2	T
1	2	4	0	57	1	T
0	3	14	1	57	1	T
1	1	1	0	58	1	P
1	4	24	0	57	2	E
0	3	14	1	56	1	E
1	1	1	0	57	9	P
1	1	24	0	57	1	T

health, 9-stands for unknown. Gender code 1 means male, 0-female. The variable Data_Frame ensures the random distribution of the database records to Training-T, Evaluation-E and Prediction-P sets.

The logit regression model can be processed in BigQuery ML, here we need to open Google Cloud platform, BigQuery sandbox, set a new project, create the dataset and upload the data file. Designing the logistic regression model consists of the following steps:

1. Create and train the logistic regression model on training data.
2. Evaluate the model performance with evaluation set of data
3. Predict the output from inputs prediction data

For model creation task we can write a simple SQL query (Fig. 10.3).

Here the *'Logit.Logit'* is the name assigned to uploaded table. The achieved performance of logit regression by classifying *Vital_Status* can be seen from Fig. 10.4.

In Fig. 10.4, the confusion matrix is presented as a table in which predictions are represented in columns and actual status is represented by rows. The performance of the model is explored by applying several characteristics of precision evaluation:

Table 10.3 Codes for treatment methods

Code	Anticoagulation	Dual antiplatelet therapy	Thrombolysis	Others
1	x			
2		x		
3			x	
4				x
12	x	x		
13	x		x	
14	x			x
23		x	x	
24		x		x
34			x	x
123	x	x	x	
134	x		x	x
234		x	x	x
0				

👤 Logit_create Edited

```
1  CREATE OR REPLACE MODEL
2    `Logit.Logit_model`
3  OPTIONS
4    (model_type='LOGISTIC_REG', auto_class_weights=TRUE, input_label_cols=['Vital_Status'])
5  AS
6  SELECT * EXCEPT(Data_Frame) FROM `Logit.Logit` WHERE Data_Frame = 'T'
```

Fig. 10.3 Model creation statements

Score threshold   Confusion matrix

Positive class threshold	——⬤—— 0.4696
Positive class	1
Negative class	0
Precision	0.7732
Recall	0.7683
Accuracy	0.6956
F1 score	0.7707

Actual labels / Predicted labels

Actual labels	1	0
1	76.83%	23.17%
0	44.94%	55.06%

Fig. 10.4 Evaluation of trained logit model

accuracy, recall and precision. The characteristics of *Accuracy* shows what percent of all values are correctly predicted by the model. In our case the general accuracy of prediction is close to 70%. *Recall* calculates the percent of correct predictions of *Vital_Status* for all the true values (=1). It means, that the performance of the model for the Vital Status value "1" is better, than general performance, and equals to 76,83%. The proportion of the instances which were correctly recognized as positive (per total positive predictions) is called the *Precision*. The *F1 score* denotes the harmonic mean of *Precision* and *Recall*. The accuracy of the model may be satisfactory or not sufficient depending on the requirements and complexity of the task solved. In the Fig. 10.4 shows performance of the Logit regression model on training set.

To see the performance of our model on evaluation set we can write an evaluation SQL query on evaluation set (Fig. 10.5).

When the SQL query is executed, BigQuery calculates the accuracy and other model performance characteristics on evaluation set (Fig. 10.6).

As we can see the logit regression model performs on evaluation set even better than on training set. The accuracy and other ratio estimates shows good classification power of *Vital_Status* variable.

The last logit regression modelling step provides model adaptation for prediction set. For this case we need to write prediction SQL query (Fig. 10.7).

The execution of this query let us find the predictions of *Vital_Status* and present the model application results in table (Table 10.4).

👤 Logit_Evaluate Edited

```
1  SELECT
2    *
3  FROM
4    ML.EVALUATE (MODEL `Logit.Logit_model`,
5    (SELECT * FROM `Logit.Logit` WHERE Data_Frame = 'E'))
```

Fig. 10.5 SQL for model evaluation

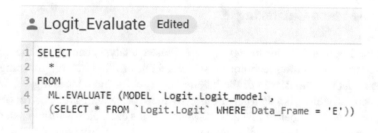

Row	precision	recall	accuracy	f1_score
1	0.8041958041958042	0.7718120805369127	0.7268722466960352	0.7876712328767124

Query complete (0.9 sec elapsed, 47.1 KB processed)

Fig. 10.6 Evaluation results

👤 Logit_predict Edited

```
1  SELECT
2    *
3    EXCEPT(predicted_Vital_Status_probs)
4  FROM
5  ML.PREDICT (MODEL `Logit.Logit_model`,
6    (SELECT * FROM `Logit.Logit` WHERE Data_Frame = 'P'))
```

Fig. 10.7 SQL query for prediction set results

Comparison of the columns "Predicted_Vital _status" and the original values "Vital_status" in Table 10.4 shows that part of the predicted values differ from the original ones, but the overall accuracy calculated for all Prediction set records is equal to 0.684426. This value lets us to conclude the good logistical classification capabilities by applying this method. The presentation of the outcomes in Table 10.4 gives possibility for the advanced further analysis, as the expert analysis of the incorrectly predicted cases may bring insights for adding more input variables, or introduce changes to their coding in order to reduce confusion among the predicted classes and increase the accuracy of model.

10.4 k-Means Clustering

Unlike the Logit and Probit modelling, the Cluster analysis belongs to the class of the unsupervised learning techniques, which enables to find natural groupings and patterns in data, without need of the labelled data set for model training.

K-means clustering is a data partitioning method for assigning records (or objects) to the predefined number of clusters. K-Means treats each observation as an object that has its location in a multidimensional space. The algorithm of k-Means finds a partition in which objects within each cluster are as close to each other as possible, and, at the same time, as far as possible from the objects of the other clusters. Based on the attributes of our data, we can select one of the generally applied distance metric to be used by the k-Means model for calculating distances among the clusters and distances between the instances within cluster.

As k-Means clustering creates a single level of clusters it is suitable for both large amounts of data objects and numerous attributes. Each cluster in a k-Means partition consists of its member objects and has a predefined centre or centroid. K-Means method tries to minimize the sum of the distances between the centroid and each member object of the cluster. The computation procedure depends on the applied distance metrics. By default, k-Means uses the squared Euclidean distance metrics to determine distances. The visualisation of the output of the method plots

Table 10.4 Prediction results for prediction set

Query complete (0.3 s elapsed, 47.1 KB processed)

Row	Predicted_vital_status	Vital_status	Stroke_type	Treatment_methods	Health_status	Age	Gender	Data_frame
101	1	1	1	14	0	63	1	P
102	0	0	3	14	1	61	2	P
103	1	1	1	14	0	61	2	P
104	1	0	1	14	1	61	2	P
105	0	1	3	14	1	53	1	P
106	1	1	3	14	0	50	1	P
107	1	1	3	14	1	48	1	P
108	1	1	3	14	2	44	2	P
109	0	1	1	23	0	86	2	P
110	1	0	1	23	1	76	2	P
111	1	1	1	23	2	74	1	P

the clusters on the two-dimensional space for simplification of the analysis, however the underlying computations deal with the multidimensional settings.

The following steps are performed for k-Means clustering (k-Means Clustering): [14]:

1. Examine k-Means clustering solutions for different selected number of clusters k to determine optimal number of clusters for the data set. Some tools (such as Statistica or Viscovery SOmine) offer estimation of optimal number of clusters;
2. Evaluate clustering solutions by analysing silhouette plots and silhouette values, or based on criteria, such as Davies–Bouldin index values, and Calinski–Harabasz index values;
3. Replicate clustering from different randomly selected centroids and return the final solution with the lowest total sum of distances among all the replicates.

A silhouette value is a standard measure of how close the points of one cluster are to the points of the adjacent clusters. This measure takes values from interval $[-1,1]$. The value "-1" denotes the points that are probably assigned to the wrong cluster, and silhouette value equal to 1 indicates points that are very distant from the neighbouring clusters. Usually silhouette values are presented graphically by the silhouette plot, which enables to choose the right number of clusters.

The criteria-based method for finding the optimal number of clusters include calculation of Davies–Bouldin (Davies–Bouldin index) or Calinski–Harabasz (Indice de Calinski–Harabasz) index values. Without going into the technical details of calculating these indicators, we will summarize that the optimal number of clusters is indicated by the lowest indicators values.

To illustrate the application of the k-Mean clustering method, we used the extended version of the previously described data file containing various health and personal characteristics of patients diagnosed with stroke. Part of the attributes of this file were explained in Table 10.1, such as the variables of Vital_Status, Type of stroke, Treatment methods, Health modified ranking score before stroke, Age at stroke, Gender, and Days spent in hospital.

For the further study the dataset was expanded by variables expressing other characteristics of the patient history with the assumption that additional knowledge about the patient may increase prediction power of the mode. The information whether there was a stroke before, specific Stroke symptoms, and indication of Health complications may enable us to better distribute the patients into meaningful groups and recognise the useful patterns of data. The example of data file records used for k-Means clustering is presented in Table 10.5. According to the concept of multidimensional space associated with clustering computational models we can imagine the file records as points in 9-dimensional space. Contrarily to the supervised methods, all variables serve as inputs.

For this example, we filtered only the record of patients with the Vital_status $= 1$ (Alive), therefore 642 records we used for research of K-Means clustering. The clustering of patients into predefined number of clusters can be useful in case of meaningful categories (cluster) applied in medical practice, such as separating

Table 10.5 Example data file records for clustering

Days at hospital	Stroke_type	Treatment_methods	Health_status	Age	Gender	Past_stroke	Stroke_symptoms	Health_complications
8	3	0	1	17	2	0	0	0
23	3	4	2	16	1	0	12	0
6	1	24	0	91	1	0	2	0
16	1	24	0	91	1	0	13	0
18	3	4	0	92	1	0	1	0
21	1	24	4	91	2	1	23	0
8	1	2	3	90	2	1	2	0
8	1	24	0	88	1	0	23	0
2	1	24	0	88	2	0	2	0
6	1	24	0	88	2	0	2	3
7	1	24	0	88	1	0	23	4
8	1	4	0	87	1	0	23	2
14	1	4	0	86	2	0	12	0
7	1	24	9	88	2	0	123	0
22	1	24	0	86	2	0	23	0
6	1	12	2	86	2	1	123	0
13	1	4	0	87	2	0	13	4

patients for Rehabilitation, Medication prescription or for appointment of specific health strengthening procedure.

In Table 10.5, *Past_Stroke* value equal to 1 means the repeated stroke case, *Stroke_Symptoms* can take values of: 0-No symptoms, 1-Impaired consciousness, 2-Weakness/paresis, 3-Speech disorder (aphasia) or joint occurrence of several symptoms, e.g. 13 indicates Impaired consciousness and Speech disorder (aphasia). *Health_complications* are divided into four different groups: 1-deep vein thrombosis, 2-other CV complications, 3-pneumonia, 4-other complications. 0 value stands for unspecified complications, and, similarly, the code 13 expresses the double complications of deep vein thrombosis and pneumonia.

The k-Mean clustering can be done by various software, but here we will use MATLAB R2020b version. MATLAB® [15] combines a desktop environment tuned for iterative analysis and design processes with a programming language that expresses matrix and array mathematics directly. Using the predefined Matlab functions we can perform all popular classification, regression, and clustering algorithms for supervised and unsupervised learning.

Matlab enables us to fine tune all parameters of clustering by writing a program code, and to find the optimal number of clusters, as well as to evaluate the clustering solutions by analysing silhouette values, Davies–Bouldin and Calinski–Harabasz index values.

The analysis and clustering of the described data file by using k-Mean algorithm is executed by the Matlab program code presented in Fig. 10.8.

The operator on program line 6 enables us to specify a testing set for the unsupervised learning of k-Means algorithm and to select the attributes. As it it specified by operator 6, for this case we have selected 440 cases starting from record 21 to 460. After initial computation phase we have noticed that variables Days at hospital and Gender have negative influence to the K-Means performance. So, for the following

```
Editor - C:\MATLAB7\work\Kmeans.m
   Kmeans.m  ×  +
1 -    clc;
2 -    clear;
3      %Virgilijus(c) 2021
4 -    load('K_Meansl.mat')% Loading the data file
5 -    disp(size(M)); %display the number of records and atributes
6 -    M=M(21:460,[2 3 4 5 7 8 9]);% selecting testing set and attributes
7 -    CH=evalclusters(M,'kmeans','CalinskiHarabasz','KList',[1:5]);%Davies-Bouldin index values
8 -    DB=evalclusters(M,'kmeans','DaviesBouldin','Klist',[1:5]);%Calinski-Harabasz index values
9 -    disp(CH);disp(DB)
10 -   [idx,C,sum] = kmeans(M,2,'Distance','cityblock','Display','Iter'); % K-Means function
11 -   [silh,h] = silhouette(M,idx,'cityblock'); %Calculation of silhouette value
12 -   xlabel('Silhouette Value') %silhouette plot
13 -   ylabel('Cluster')
14 -   D = mean(silh);
15 -   fprintf('Average silhouette values = ');disp(D)
16 -   disp(C) % returns the k cluster centroid locations
17 -   disp(sum) % sums of distances to centroid centre
18     %idx(41:50) % Display the clusters for 41:50 records
19 -   disp(nnz(idx==1));disp(nnz(idx==2));disp(nnz(idx==3))%the number of cases in clusters
```

Fig. 10.8 Matlab code for k-Means algorithm

```
CalinskiHarabaszEvaluation with properties:

    NumObservations: 440
          InspectedK: [1 2 3 4 5]
    CriterionValues: [NaN 4.8761e+03 1.1386e+04 9.8071e+03 8.3579e+03]
           OptimalK: 3

DaviesBouldinEvaluation with properties:

    NumObservations: 440
          InspectedK: [1 2 3 4 5]
    CriterionValues: [NaN 0.0675 0.1944 0.3741 0.5701]
           OptimalK: 2
```

Fig. 10.9 Calinski–Harabasz and Davies–Bouldin criterion values

Fig. 10.10 k-Means accuracy verification for 3 clusters

```
    iter    phase          num                sum
     1        1            440              10755
Best total sum of distances = 10755
Average silhouette values =      0.7695
```

computation stage we excluded them from our research, and tried to find the clusters only by selecting 2, 3, 4, 5, 7, 8, 9 attributes (see Table 10.4).

In order to find the optimal number of clusters we calculated the Davies–Bouldin and Calinski–Harabasz index values (8 and 9 program lines, Fig. 10.8). The output of 9 line is presented in Fig. 10.9.

The optimal number of 3 clusters was suggested by the Calinski–Harabasz criterion, but Davies–Bouldin criterion advices optimal number of 2 clusters. Therefore we explored both cases of 3 and 2 clusters for evaluation.

The estimation of the K_Mean model to our data was started with k = 3 (line 10), it calculated the best total sum of distances to the centroids and average silhouette values. The calculation results are presented on Fig. 10.10.

In Fig. 10.10 the silhouette values equal to 0.7695, which confirm the excellent partition of our cases to 3 clusters. The silhouette plot on Fig. 10.11 visually confirm this assertion. Only very small number of cases have silhouette values less than 0.6 (Fig. 10.11).

Application of k-Means model calculations for 2 clusters show worse performance comparing to the case of 3 clusters (Fig. 10.12).

Although the average of silhouette values of 0.6846 for 2 clusters only differ by small amount from those of 3 clusters. However, the selection of partition of cases to 3 clusters may be more adequate by final expert judgement. After applying the selected K-means clustering model, the clusters can be further explored according to numerous characterstics of the variables included to different clusters. We use Matlab program code to calculate the Number of cases in clusters (line 19) and Sums

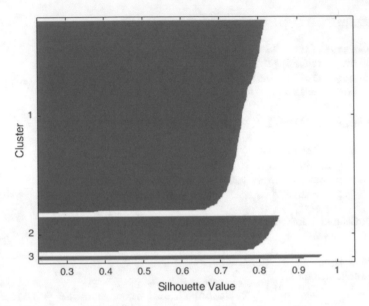

Fig. 10.11 The silhouette plot for 3 clusters

Fig. 10.12 k-Means
accuracy verification for 2
clusters

```
iter    phase        num              sum
 1        1          440            19032
Best total sum of distances = 19032
Average silhouette values =      0.6846
```

of distances to centroid centre (line 17) in order to characterize the size and similarity of objects within clusters (Table 10.6).

In order to check the membership of a particular patient (or group of patients) to some cluster we may apply different functions of the machine learning environment, such as Matlab: as an example, the command in line 18 (Fig. 10.8) displays the clusters for cases from 41 to 50.

Based on the demonstrated example, we can state that the application of Matlab for machine learning algorithms has a high degree of configuration freedom, allows the researcher to control the parameters of the method and test various computational scenarios. Understanding the background principles of the machine learning

Table 10.6 Cluster
information

k-Means clusters (k = 3)	Number of cases in clusters	Sums of distances to centroid centre
1	366	9188
2	67	1266
3	7	301

models and the flexibility of their application in different computational environments enables domain experts and researchers to derive important analytical insights.

10.5   Artificial Neural Network

The Artificial Neural Network (ANN) model is inspired by the biological neural network. It can learn to perform tasks by observing examples, without applying any rules of a particular task.

The ANN model and its modifications is widely used in various application domains, such as language recognition, machine translation models, social network filtering, facial recognition, financial instrument prediction, and many more, where the tasks of classification or time series forecasting are relevant. In medicine, ANN is used to diagnose various diseases and their complications, to evaluate the effects of drugs, to predict the duration of treatment, or to cluster medical anomalies. ANN may link the symptoms of patients with a specific disease, and learn to identify the disease accordingly.

Contrarily to the statistical methods, the ANN is a data-driven approach, therefore the ANN model is trained by available data set for applying it in the testing conditions of the researched domain. For each of the tasks, it is necessary to set up an appropriate neural network. The following methodology should be followed:

1. Preparation of data for the study. These include data collection, organization, normalization, preparing the training and testing sampling.
2. Selection of ANN structure. It is determined by the number of outputs, input variables, hidden layers and the number of neurons of the model. The neuron connection principles, threshold and transmission functions should be determined as well.
3. ANN training. The network training strategy, training algorithm and the training effectiveness needs to be evaluated.
4. Network testing. The evaluation of the created neural network is performed by using an input data set, other than the one used for its training.

All these tasks are highly interrelated and influence the quality of the model. Depending on the available input data set and the task being solved, the appropriate network structure is modelled for applying the most suitable ANN training algorithm. The two most common neural network structures, such as Single-layer perceptron and backpropagation network (multilayer perceptron) are further discussed and explored by presenting the experimental sample.

Single-layer perceptron

Single-layer perceptron is the simplest form of ANN used to classify linearly separated structures. It is a single-layer direct propagation neural network with a threshold transmission function (Fig. 10.13). Rosenblat [16] proved that if such a network is

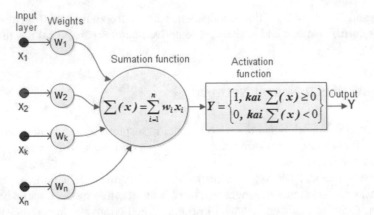

Fig. 10.13 Single-layer perceptron

trained by examples from linearly separable classes, then the perceptron algorithm converges and finds a hyperplane separating those classes.

The solution of the perceptron equation $\sum(x) = 0$ defines a line or hyperplane as the boundary between distinct classes. The solution is obtained by learning the network and choosing the correct network weights. As mentioned, perceptron can only distinguish between linearly separable classes (Fig. 10.14). To describe the structure and training algorithm of perceptron, we will use notations according to Hajek [17].

Perceptron learning is a supervised learning system. Thus the training set consists of pairs $(x^{(p)}, d^{(p)})\big|_{p=1}^{N}$, where $x^{(p)}$ denote the input vector $x^{(p)} = (x_1^{(p)}, x_2^{(p)}, \ldots, x_n^{(p)})^T$, and $d^{(p)}$ is the known output a vector (teacher) whose components can acquire only two values: 0 or 1. Let $y^{(p)}$ be the output vector of the neural network.

The error function can be introduced as a vector:

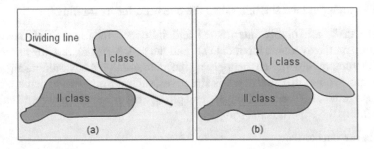

Fig. 10.14 Linearly separable classes (**a**), Not separable linearly classes (**b**)

$$J = \sum_{p=1}^{N} \left(y^{(p)} - d^{(p)} \right) w^{(p)} x^{(p)}$$

The neural network correctly separates classes when $J = 0$. In all other cases, the separating plane is not found.

In the practical implementation of perceptron training, we change the weights according to the given formula until J becomes as small as possible and no longer changes. If $J = 0$, then our classes were linearly separable and we separated them. If $J \neq 0$, then the classes were not linearly separable and we found the most appropriate separation of those classes.

Several perceptrons can be combined into a more complex network. Such a structure makes it possible to distinguish more complex classes of objects, such as those that can be separated by a plane or a hyper polygon. Figure 10.15 shows a perceptron network with many input and output neurons.

As the perceptron neural network consists of individual perceptron's, each of those can be trained separately according to the algorithm described above. In the 1960s, when perceptron networks became very popular, many researchers thought that any intelligent systems could be constructed with the help of perceptron networks. Unfortunately, it later turned out that far from all systems are so simple. When in 1986 elementary McCulloch-Pitts neural networks was replaced by networks with differentiated activation function and an advanced backpropagation algorithm was described, many complicated systems could be modelled by using such neural networks [18].

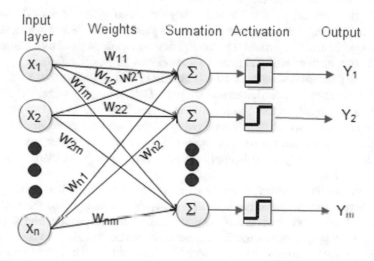

Fig. 10.15 Perceptron network

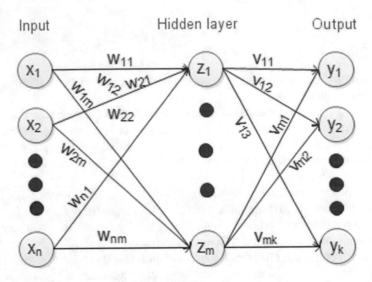

Fig. 10.16 Multilayer perceptron with one hidden layer

Backpropagation networks

The backpropagation network is often referred as the direct propagation multilayer perceptron (Fig. 10.16). His training run with the teacher, having a test set, and the teaching algorithm is called a backpropagation algorithm using a gradient descent method to minimize the total squared error.

The backpropagation algorithm was firstly described in the work of Bryson and Yu-Chi Ho [19], but it did not receive wider recognition until 1986, when Rumelhart et al. [18] published their article. The later period was characterized by a particularly strong development of artificial neural networks and their application.

Using the gradient descent method, it is necessary to differentiate the transfer function with respect to input variables and weights. Thus, nonlinear sigmoidal function or hyperbolic tangent is most commonly used in backpropagation networks. Multilayer perceptron allows the classification of more than just linearly separable classes. Depending on how many neurons are in the hidden layer, we can obtain a separation surface as a convex polygon with approximately as many edges as there are neurons in the second layer.

Once the ANN topology is established, we need to adapt the training algorithm, where a backpropagation algorithm is applied for training of the multilayer perceptron. It consists of two phases: propagation forward and propagation backward.

As the ANN propagates forward, the input variables are transformed layer by layer into output layer variables using fixed weights, thresholds, and transfer functions. In the backpropagation phase, all network weights are recalculated depending on the size of the error signal, which is calculated as the difference between the values of the ANN output variables and the predetermined output vector (teacher).

The opportunity to learn from examples and gain experience has allowed neural networks to be widely used to solve practical problems. Artificial neural networks can help to examine the structure of data, determine its trend, make a forecast, assess risk, or predict impending anomalies. To do this, the neural network must be trained using historical data. The ANN is most commonly used to address classification or clustering challenges because of its greater accuracy and flexibility compared to traditional statistical methods. The most critical challenge for application of the ANN principle in healthcare and other high risk decision domains lays in its "black box" structure: as the model learns from the data set with the labelled output (a teacher) it learns how to estimate the output from the input variables, but it does not provide rules or formulas for clarifying dependencies for decision making. Numerous modifications of the neural network algorithms are proposed in the research works on different conceptual development areas for creating transparency of the ANN performance.

The experimental research

The selected classification task concerns rehabilitation assignment for the patients who have experienced stroke. The experimental analysis was performed for the input data set presented in Table 10.5. As the neural network is a supervised learning algorithm we needed the output variable for training and testing the best performing NN model. Therefore, one more variable of the historical stroke patient database was included, which denotes rehabilitation type prescribed by the expert doctors during the hospitalization. There were four types of rehabilitation therapy (Table 10.7), and the cases with no assignment for rehabilitation were excluded.

Successful solving of this task leads to creating a neural network model which could forecast the output: propose relevant rehabilitation type according to the health characteristics of the patient. The model could also serve to better plan human resources, as different types of rehabilitation required involvement of different specialists and schedule their time.

The analysis enables to solve what kind of rehabilitation is most likely to be prescribed according to the nine input variables (Table 10.5) serving as health characteristics of the patient.

Several experiments were performed for exploring ANN models performance. In the first step, the neural network models were generated from the data set by applying different algorithms, and three best performing models were retained for

Table 10.7 Rehabilitation types

Code	Code value	Rehabilitation prescribed
1	RWt	Working therapy
2	RPt	Physical Therapy
3	RSp	Speaking exercise
4	RSw	Swallowing exercise
0	RNo	No rehabilitation

| Index | Model Summary Report (DataRehabNN Rh) | | | | | | |
	Profile	Train Perf.	Select Perf.	Test Perf.	Inputs	Hidden(1)	Hidden(2)
1	Single Layer NN 9:39-4:1	0,84	0,85	0,79	9	0	0
2	MLP 10:45-12-11-4:1	0,95	0,81	0,76	10	12	11
3	RBF 9:44-17-4:1	0,83	0,88	0,81	9	17	0

Fig. 10.17 Multilayer perceptron with one hidden layer

further analysis. The second step had to explore the accuracy of the models in solving the classification task by analysis their general performance, and the confusion among the output classes. The third step had to reveal the importance of different variables for building the neural network model. The last step had to investigate classification behaviour at different value ranges of the variables. The last two steps had to provide solution for the "black box" nature of the ANN models. In the healthcare problems the situation of the "black box" is mainly not acceptable, as it means that the ANN model may just advice the output without providing rules or explanatory insights, therefore many modifications and solutions of the ANN algorithms are being explored for converting ANN to "grey box" or "white box".

The STATISICA for Windows software was used to design the neural network. The data set was randomly split to three subsets used for training (70%), selecting evaluation set (15%), and testing (15%).

Three models were retained (Fig. 10.17), we can see that different algorithms, such as Single Layer perceptron NN, MLP (multilayer perceptron), RBF (Radial basis function), had similar performance. The MLP model was the most accurate in the training stage (0,95), whereas the RBF was slightly better in the testing stage, which may indicate good performance for the unknown new data set. The general classification precision of the models in different stages varied between 0,79 and 0,95 (Fig. 10.17), which indicates good possibility to propose most suitable rehabilitation type. The structure of the neural network models retained is described by their profile data (Fig. 10.17), which denotes number of input variables (9), number of neurons in each hidden layer, and one output variable with the four classification outcomes (Table 10.7).

As the model aims to correctly select the output value, namely the rehabilitation type, we may explore the performance of different models while assigning particular output values. In Fig. 10.18 the confusion matrix reveals, that the Single Layer NN model had quite significant confusion among classes: it could not assign the rehabilitation types of RWt and RSp to any of the classes, while most of the cases of RSw were wrongly assigned to RPt. Similar confusion problems were demonstrated by the RBF model. Despite similar general accuracy of the models, the best ability to recognize different output classes was shown by MLP model.

The confusion problem may be determined by different number of cases with various output, used for training the models. In our case the biggest number of cases had the output variable value RPt; or it may be determined the significance of different variables which may be explored by sensitivity analysis of the designed

	Confusion Matrix - RehabA(1,2,3) (DataRehabNN_Rh)			
	RWt	RPt	RSw	RSp
RWt.1-Single Layer NN	0	0	0	0
RPt.1-Single Layer NN	37	520	31	23
RSw.1-Single Layer NN	2	16	12	1
RSp.1-Single Layer NN	0	0	0	0
RWt.2- MLP	12	4	0	0
RPt.2- MLP	23	507	10	13
RSw.2- MLP	4	17	32	4
RSp.2- MLP	0	8	1	7
RWt.3- RBF	0	0	0	0
RPt.3- RBF	39	536	43	24
RSw.3- RBF	0	0	0	0
RSp.3- RBF	0	0	0	0

Fig. 10.18 Confusion matrix

	Sensitivity Analysis - Models 1, 2 and 3 (DataRehabNN_Rh)					
	Days at Hospital	Stroke_Type	Treatment_methods	Age	Stroke_Symptoms	Health_complications
Ratio.1:Single Layer NN	1,00	1,00	0,00	1,00	1,00	1,00
Rank.1:Single Layer NN	6	2	9	3	8	1
Ratio.2: MLP	1,01	1,11	1,11	1,26	1,23	1,05
Rank.2: MLP	8	4	3	1	2	5
Ratio.3: RBF	1,00	1,02	1,02	1,03	1,03	1,01
Rank.3: RBF	7	4	3	1	2	5

Fig. 10.19 Sensitivity analysis of the most influential variables

neural network models. In Fig. 10.19 the variables are ranked by calculating ratio of their significance.

In Fig. 10.19 the sensitivity analysis revealed different importance of the variables for generating ANN models. The most influential variables are shown: for the Single Layer NN the *Health complications* were ranked 1st, but for the MLP and RBF models the *Age* and *Stroke Symptoms* were ranked correspondingly 1st and 2nd. The sensitivity analysis may advice the areas for more detailed investigation and improving precision of the models. It can be achieved by enhancing richness of data in the areas related to the most significant influences and identifying most vulnerable areas of inaccurate performance.

The performance of the MLP model in recognizing values of the output variables denotes strongest reliability of the model for rehabilitation of type RPt (94,6% correct), while the model is not useful for the RWt (30,77% correct) and RSp (29,17% correct). It can be noticed, that relatively small number of cases with different outputs is not the determining factor, as the RSw (43 cases) accuracy is 74,42% whereas RWt had similar number of cases (39) with much lower performance (30,77%) (Fig. 10.20).

Application of the ANN algorithms and models in healthcare has broad potential due to their computational power, and as the regression, classification or time series-related tasks are important in the healthcare processes related to diagnosis, treatment, rehabilitation and many others. However, the experimental research has demonstrated necessity to apply various approaches not only for building models and analysing

	Classification (MLP) (DataRehabNN_Rh)			
	RehabA.RWt	RehabA.RPt	RehabA.RSw	RehabA.RSp
Total	39	536	43	24
Correct	12	507	32	7
Wrong	27	29	11	17
Unknown	0	0	0	0
Correct(%)	30,77	94,6	74,42	29,17
Wrong(%)	69,23	5,4	25,58	70,83
Unknown(%)	0,00	0,0	0,00	0,00

Fig. 10.20 Sensitivity analysis of the most influential variables

their general accuracy, but for their in-depth analysis of performance, influences and possible sources of vulnerabilities and inaccuracies.

10.6 Conclusion

The chapter provides essential characteristics of methods, traditionally applied for data processing, such as regression analysis, as well as their modifications towards the area of artificial intelligence methods, such as logit, probit models, K-means, Neural networks. The healthcare domain uses variety of data sources and measurement scales, as well as different target requirements for output information. It implies that different methods have to be considered for solving tasks, while the in-depth analysis of the generated solution models may bring to adoption or rejection of different models due to their imbalanced reliability in different classes, segments of cases. The performance of the methods, their analytical power and relevance to the healthcare application domain is illustrated by brief experimental computations for investigation of stroke patient database. Various software tools, such as STATISTICA, Matlab, Google BigQuery ML were applied for analysis, ensuring broad variety of analytical tools for in-depth analysis of generated solutions and deriving new insights for their improvement. The regression analysis, characteristics and the experimental examples of their applications reveal advantages, disadvantages, and causes of irrelevant application of the methods. The analytical tools not only enhance transparency of the artificial intelligence data driven models, but may indicate areas of improving data quality, or initiate potential sources for supplementing enriched data related to the most influential variables characterizing persons and various aspects of healthcare.

Acknowledgements This publication is based upon work from COST Action "**European Network for cost containment and improved quality of health care-CostCares**" (CA15222), supported by COST (European Cooperation in Science and Technology)

COST (European Cooperation in Science and Technology) is a funding agency for research and innovation networks. Our Actions help connect research initiatives across Europe and enable scientists to grow their ideas by sharing them with their peers. This boosts their research, career and innovation.

https://www.cost.eu

References

1. Buch V.H., Ahmed, I., Maruthappu, M.: Artificial intelligence in medicine: current trends and future possibilities. Br. J. Gen. Pract. **68**, 143–144 (2018)
2. Hosny, A., Aerts, H.J.W.L.: Artificial intelligence for global health. Science **366**, 955–9556 (2019)
3. Neill D.B.: Using artificial intelligence to improve hospital inpatient care. IEEE Intell. Syst. **28**, 92–95
4. Panch, T., Pearson-Stuttard, J., Greaves, F., Atun R.: Artificial intelligence: opportunities and risks for public health. Lancet Dig. Health **1**, e13–e14 (2019)
5. Reddy, S.: Use of artificial intelligence in healthcare delivery, In: eHealth-Making Health Care Smarter, pp. 81–97. IntechOpen (2018)
6. Sanders S.F., Terwiesch, M., Gordon, W.J., et al.: How Artificial Intelligence Is Changing Health Care Delivery. https://catalyst.nejm.org/health-care-aisystems-changing-delivery/. Accessed 10 Jan 2021
7. Triantafyllidis A.K., Tsanas, A.: Applications of machine learning in real-life digital health interventions: review of the literature. J. Med. Internet Res. **21**, e12286 (2019)
8. BigQuery: https://cloud.google.com/bigquery-ml/docs/introduction Accessed 10 Jan 2021
9. Davies–Bouldin index: https://en.wikipedia.org/wiki/Davies%E2%80%93Bouldin_index. Accessed 10 Jan 2021
10. Nisbet R., Elder J., Miner G.: Statistical Analysis & Data Mining Applications. Elsevier, Canada (2009)
11. Statistica: http://www.statsoft.com/textbook/. Accessed 10 Jan 2021
12. Krikščiūnienė, D., Sakalauskas, V.: Intelektiniai modeliai marketingo sistemose: Monograph in Lithuanian. Vilnius, Vilniaus universiteto leidykla, 384 p (2014)
13. Pfaffenberger Roger, C.: Patterson Jamer H. Statistical methods for business and economics.-Richard D. Irvin, INC., 828 s. (1981)
14. k-Means Clustering: https://se.mathworks.com/help/stats/k-means-clustering.html. Accessed 10 Jan 2021
15. Matlab: https://se.mathworks.com/products/matlab.html?s_tid=hp_products_matlab. Accessed 10 Jan 2021
16. Rosenblatt, F.: The perceptron: A probabilistic model for information storage and organization in the brain, cornell aeronautical laboratory, Psychological Review, vol. **65**, No. 6, 386–408 (1958). https://doi.org/10.1037/h0042519
17. Hajek M.: "Neural Networks". (2005). http://www.cs.unp.ac.za/notes/NeuralNetworks2005.pdf
18. Rumelhart, D.E., Hinton, GE., Williams, R.J.: Learning representations by back propagating errors, Nature **323**, 533–536 (1986)
19. Bryson, A.E.Jr., Yu Chi Ho.: Aplied optimal control. Blaisdell Publishing Co. (1969)
20. Indice de Calinski-Harabasz: https://fr.wikipedia.org/wiki/Indice_de_Calinski-Harabasz. Accessed 10 Jan 2021

Chapter 11
Probabilistic Modelling and Decision Support in Personalized Medicine

Michal Javorník, Otto Dostál, and Aleš Roček

Abstract The concept of personalized medicine, often called the biggest revolution in medicine, is becoming an emerging practice. The article presents personalized medicine in a broader context as an interdisciplinary issue covering the current trends of information and communication technology in medicine, legal aspects, and probabilistic network modelling. Employing the concept of probabilistic network reasoning means extracting the meaningful knowledge, mathematizing it, incorporating the particular patient information and then using inference mechanisms of the created mathematical model for personalized decision support. Bayesian networks can serve as a multidimensional decision support framework representing the real-world medical domain. Their power, together with the possibilities of global sharing of necessary medical knowledge, represents a promising approach of extracting new, often hidden, knowledge about the given medical domain and thus opens up new ways of achieving the delivery of personalized medicine. Establishing patient diagnosis and treatment prognoses are the critical issues in personalized decision support. Mathematical modelling is beginning to play an irreplaceable role here.

Keywords Bayesian networks · Decision support · Personalized medicine · Probabilistic modelling

M. Javorník (✉) · O. Dostál · A. Roček
Masaryk University, Brno, Czech Republic
e-mail: 1111@mail.muni.cz

O. Dostál
e-mail: 45081@mail.muni.cz

A. Roček
e-mail: 205054@mail.muni.cz

© The Author(s) 2022
D. Kriksciuniene and V. Sakalauskas (eds.), *Intelligent Systems for Sustainable Person-Centered Healthcare*, Intelligent Systems Reference Library 205,
https://doi.org/10.1007/978-3-030-79353-1_11

11.1 Introduction

The concept of personalized medicine, often called the biggest revolution in medicine, is becoming an emerging practice. Turning it into an individualized approach means using specific information about a patient's health condition to make a diagnosis, plan a treatment, or establish a prognosis.

This article presents personalized medicine in a broader context as an interdisciplinary issue covering the current trends of information and communication technology in medical imaging, legal aspects, and probabilistic network modelling.

Probabilistic network modelling [1] represents one of the most promising approaches. Nowadays, it can build on a broad spectrum of currently available knowledge reflecting the examined medical domain. It can be the specific knowledge of medical experts, known general principles of the domain/disease, detailed patient data, knowledge hidden in the vast digital medical datasets, etc. As medical data are mostly collected during daily routines and not as a result of coordinated research activities, the process of collecting and extracting the datasets necessary to build and evaluate the model becomes more difficult.

In short, employing the concept of probabilistic network reasoning means extracting the meaningful knowledge, mathematizing it, incorporating the particular patient information and then using inference mechanisms of the created mathematical model for personalized decision support. In practice, the following initial steps must be performed:

- development of a framework enabling the efficient collection of relevant experts' opinions and knowledge from clinical practice about the cases close to the disease under study;
- construction of a probabilistic graphical model via a probabilistic machine learning approach or manually based on explicit medical expert knowledge if there is a lack of data;
- the expression of the initial health state of a particular patient via a probabilistic distribution.

Probabilistic graphical models are used to answer complex questions that are difficult to solve using traditional probabilistic approaches. One of these models' critical features is their explanatory capabilities, which are essential for discussions with domain experts in the phase of model construction and when communicating computed results.

An accurate understanding of relevant human biological mechanisms, interoperability principals in the healthcare domain, and correct understanding of mathematical modelling with its inference capabilities stand behind the successful implementation of the decision support system in the healthcare area focusing on individual patient care, i.e., diagnostic and prognostic reasoning, treatment alternatives modelling, etc.

Personalized decision support systems based on probabilistic graphical models can be integrated with computer-assisted decision support. The critical question is how to incorporate its suggestions into the decision-making process. The role of experienced radiologists will remain irreplaceable, especially in the development of relevant knowledge datasets, data labelling, and in providing additional structural knowledge necessary for model building. Traditionally, experienced radiologists label medical case studies comprising learning material for students in medical faculties or young radiologists in hospitals. The same content can, in principle, serve as reference case studies when categorizing the imaging study of the diagnosed patient.

Medical imaging plays a critical role in diagnostics. Increasing its effectiveness in rare diseases' diagnostics via the employment of appropriate deep learning methods is one of the biggest challenges these days. The regional collaboration of healthcare facilities, especially those with smaller patient populations, is necessary to exploit its potential.

A diagnostic process can be more or less risky, more or less invasive, more or less costly, etc. Appropriate use of suitable artificial intelligence methods enables a highly personalized combination of diagnostic procedures, taking into account a patient's specific health state, and achieving greater diagnostic accuracy while avoiding more invasive or other riskier methods.

Accelerating research in this area also means exploring appropriate organizational models, a framework enabling the collection of relevant information, making it available for clinical practice, and protecting patient privacy at the same time. However, a broad spectrum of related legal questions also arises. For example, who is at fault when the recommendation of the decision support system is wrong?

11.2 Related Trends in eHealth

A multidisciplinary approach is an alternative to coping with the ever-increasing complexity, dynamism, and variability of today's healthcare. Emerging scientific disciplines are often methodologically linked to the sciences based on which they were created, but their achievements inspire these sciences retrospectively. One example is medical informatics, an applied science that designs new progressive procedures for many medical problems and contributes to developing healthcare knowledge. Interdisciplinary research in healthcare focuses on the development of knowledge systems, the intelligent use of the experience stored in health databases, the development of new telemedicine technologies, and especially the improvement of diagnostic and therapeutic processes. Therefore, it is necessary to integrate IT, medical, biomedical, legal, mathematical, and economic knowledge and find opportunities for their use in the environment of medical practice, teaching, and research. The daily routine of healthcare information systems and the management of health documentation in electronic form are addressed by many national and international legal standards. From the very beginning, in addition to the limits of current information and communication technologies, it is also necessary to take into account, carefully consider, and understand the legislative restrictions so that the initial considerations respect the relevant legal norms arising from the legal system.

Today, the most progressive healthcare institutions use advanced software systems to support decision-making, sophisticated image processing techniques that make it possible to search for similar cases (evidence-based medicine) in available knowledge databases, and a whole range of methods of artificial intelligence. However, applications are often run at a mode that does not support standard medical communication protocols or does not allow the transfer and subsequent use of the information obtained outside a hospital.

The direction of further development of information systems in healthcare can be characterized by patient orientation and the possibility of global access to medical data for shared care. The goal is the facilitation and acceleration of the correct diagnosis formulation, the elimination of repeated examinations, saving the time of the patient and the doctor, etc.

Hospitals and medical research centers form a unified computerized environment these days. Besides traditional healthcare activities, they also cooperate in the area of research. For research purposes, it is necessary to have access to an extensive knowledge database of case studies [2]. Much more important than networked technologies are growing networks of medical specialists. They change their traditional thinking, cooperate on the regional level [3], share specific domain knowledge, information about their patients, etc.

11.3 Interoperability

The knowledge sharing assumes effective communication of clinical information systems of cooperating healthcare institutions. Picture Archiving and Communication Systems (PACS) with associated add-on applications providing secure sharing, communication, or other specific functionalities play a crucial role.

One of the necessary conditions for successful communication is the ability to understand the transmitted information correctly. A prerequisite for correct understanding is the structured form of the transmitted medical records related to international standards and classifications. The principle of classification is creating classes consisting of concepts that coincide in a given classification attribute. The aim of the classification is then to classify the object into a particular category. There are currently many classification systems, many terminologies, and healthcare standards, but none cover the full range of needs. The following classification systems are among the most relevant.

International Classification of Diseases (ICD) [4] was created under the auspices of the World Health Organization (WHO). According to this system, diseases are classified into 21 primary groups. Each item consists of two codes. The first code identifies the underlying disease, and the second code identifies the location or complication.

Systematized Nomenclature of Medicine (SNOMED) (IHTSDO) is a worldwide standard of clinical terminology used in IT applications in healthcare [5].

Systematized nomenclature contains more than 350,000 terms from the field of healthcare. Due to its scope, it is primarily intended for use in machine processing of medical information, electronic medical documentation, decision support, etc. The system also includes mapping to other classification schemes such as the ICD. This comprehensive classification system describes individual situations in medicine using six levels. The different levels represent a description of the case regarding topology, morphology, etiology, function, procedure, and syndrome. By combining the SNOMED and Clinical Terms classifications, the SNOMED CT (Systematized Nomenclature of Medicine Clinical Terms), currently the most complete classification in medicine, was created.

Digital Imaging and Communications in Medicine (DICOM) [6] communication standard is a comprehensive international standard for exchanging digital image data. In connection with the development of new technologies, the standard is continuously expanding. The DICOM standard includes a communication protocol and a description of the medical image data format. It defines both the structure of image data and the method of exchanging this data. The DICOM standard defines rules for coding, transmission, and storage of diagnostic descriptions of image studies.

The DICOM standard's structured report [7] consists of a hierarchically arranged structure of information objects containing text and links to pictorial and other relevant data. Each information object has a name and a unique code that allow for an accurate search. The structured report consists of a header and its content, i.e., a diagnostic description of the imaging study. The most important feature of the DICOM standard's structured report is that it is an autonomous object, independent of the information systems used by the specific healthcare institution where it is processed. The broader application of the structured report is conditioned by a change in traditional image information processing methods. Additionally, in comparison with unstructured information, it allows one to easily find and further process the required information using linked software tools. It allows for the placement of a link to relevant image data, including the reason for the reference, and to include the Presentation State object, which defines the display parameters (contrast, zoom, orientation, etc.) of a specific image. It also allows references to other information sources (previous image examinations, previous descriptions, previous measurements, etc.) being taken into account when creating the diagnostic report.

11.4   Legal Challenges

The primary major legal concern is obviously the topic of personal data. Data protection is one of the fundamental human rights. Foremost it is guaranteed by the International Covenant on Civil and Political Rights [8]. This multilateral treaty, which has been ratified by most countries in the world, deals with privacy in its article 17. According to it no one shall be subjected to arbitrary or unlawful interference with his privacy, family, home or correspondence, nor to unlawful attacks on his honor and reputation and everyone has the right to the protection of the law against such interference or attacks. In European context, currently the most crucial treaty to protect human rights and fundamental freedoms is the European Convention on Human Rights [9]. The importance of this treaty is based especially on its strong enforcement mechanism though European Court of Human Rights (which for example the Covenant on Civil and Political Rights is lacking). This convention constitutes right to respect for private and family life in its article 8. According to it everyone has the right to respect for his private and family life, his home and his correspondence. The European Court of Human Rights considers in his judicature the protection of personal data, in particular respecting the confidentiality of health data, to be a vital principle in the legal systems of all the Contracting Parties to the Convention [10]. Council of Europe is also responsible for the creation of the Convention for the Protection of Individuals with regard to Automatic Processing of Personal Data (also known as the "Convention 108"). This convention deals in greater detail with personal data undergoing automatic processing [11].

Privacy of individuals needs to be protected especially in connection with challenges posed by new technologies. Each system dealing with personal data must comply with the personal data protection legislation. In the EU, this means that the General Data Protection Regulation (GDPR) rules for such processing must be followed (Europen Union 2016). The GDPR establishes a single legal framework for the protection of personal data across the Union [12].

The GDPR defines personal data as "any information relating to an identified or identifiable natural person ('data subject')". A natural person is considered identifiable if they can be identified, in particular by reference to a certain identifier (such as a name, identification number, location data, online identifier, etc.), or by one or more factors specific to the physical, physiological, genetic, mental, economic, cultural or social identity of that natural person.

In its decisions, the European Court of Justice is interpreting this definition rather broadly. According to them, information should be considered personal data even if the identity of the natural person is not clear from the data itself, but it is possible to identify such person by combining that data with additional data that the data controller or data processor is able to obtain [13].

In the context of personalized medicine, it will first need to be determined which parts of the system involve dealing with personal data. The mathematical model itself will not be problematic on its own, though there might be personal data involved in its development. On the other hand, the usage of the model in practice will likely involve dealing with personal data, more specifically the data of the patients.

Because of the GDPR rules, processing of personal data is lawful only if one of the legitimate reasons prescribed in the GDPR is present. Determining the legitimate basis for processing is thus crucial. The controller of the system should assess and evaluate the nature, purpose, scope and context of the processing of the data, and determine the condition which best ensures its legitimacy. As we are discussing data concerning health, article 9 of the GDPR must apply.

Article 9 allow processing if the data subject has given explicit consent (letter a). However, legitimization of the processing of personal data on the basis of the consent of the persons concerned is problematic, as it might be difficult to obtain such consent, which must be free, informed, concrete and unambiguous (article 4, paragraph 11), and also any consent may be revoked by the data subject at any time. It will likely be better to consider usage of legitimate purpose under letter h, which allows processing if it is "necessary for the purposes of preventive or occupational medicine, for the assessment of the working capacity of the employee, medical diagnosis, the provision of health or social care or treatment or the management of health or social care systems and services on the basis of Union or Member State law or pursuant to contract with a health professional"; in this case data must be "processed by or under the responsibility of a professional subject to the obligation of professional secrecy under Union or Member State law or rules established by national competent bodies or by another person also subject to an obligation of secrecy under Union or Member State law or rules established by national competent bodies." Also we could consider if it is not appropriate to use the legitimate purpose of scientific research (letter j).

In our context it will thus be mainly necessary to determine whether providing personalized medicine can be considered to be part of providing healthcare. If so, it will be much easier to conclude that we have the legitimate interest required by the GDPR to process personal data. The specific details regarding what can be considered part of providing healthcare might differ according to national legislation in specific countries. In general, we believe that providing personalized medicine should be within the scope of what is covered by the legitimate purpose under Article 9, letter h, as it is directly connected to the provision of healthcare to the patients for which the medicine will be personalized.

Obviously, even if we can safely conclude that the discussed processing is lawful in general, other appropriate rules set in the GDPR must be respected, such as that it is allowed only to the extent strictly necessary and proportionate, or that the processed data must be protected.

The obligation to protect the data is written in Article 25 of the GDPR. According to the Article, the controller shall, taking into account the state of the art, the cost of implementation and the nature, scope, context and purposes of processing as well as the risks of varying likelihood and severity for rights and freedoms of natural persons posed by the processing, both at the time of the determination of the means for processing and at the time of the processing itself, implement appropriate technical and organisational measures designed to implement data-protection principles, such as data minimisation, in an effective manner and to integrate the necessary safeguards into the processing in order to meet the requirements of this regulation and protect the rights of data subjects.

One of such measures can be such as pseudonymisation. Pseudonymisation is defined in the GDPR as "processing of personal data in such a manner that the personal data can no longer be attributed to a specific data subject without the use of additional information, provided that such additional information is kept separately and is subject to technical and organisational measures to ensure that the personal data are not attributed to an identified or identifiable natural person". The management of database we are using for personalised medicine is implemented in such pseudonymised manner when possible.

Appropriate measures to protect the system from unauthorised access must be implemented both to protect personal data, and to ensure the functionality of the system. Running IT services and systems is accompanied by serious legal responsibilities. In the event of disruption of the services provided, a contractual liability might be an issue. And of course, in the event of a data breach, the legal situation is even worse. The amount and potential sensitivity of the data transferred and stored is enormous. Negligence can possibly be a valid claim. The steps the organization takes to prevent the success of the attack will impact its liability. Reasonable care by the IT professionals must be taken to protect against possible attacks [14].

Unquestionably, defending against "all" possible attacks puts enormous strain on resources. That is why risk management and evaluation are relevant. It might be impossible or unrealistic to eliminate all risks, but at least reasonably foreseeable risks should be addressed. How much effort should be spent on their prevention depends on the evaluation of their likelihood, and of the costs associated with the risk of their success compared to the costs to mitigate such an attack.

11.5 Probabilistic Modelling and Decision Support

The emerging trend of personalized medicine is closely related to rapid developments in the area of machine learning. Some of its tools are becoming highly topical as they require computational power recently virtually unattainable but currently realistically available.

Decision making, or reasoning in general, under uncertainty is a big challenge of artificial intelligence and probabilistic modelling. There is a family of probabilistic graphical models enabling reasoning under uncertainty based on strong mathematical foundations. Currently, there is a plethora of successful applications in many problem domains, including medicine. One of the most promising approaches in our area of interest are Bayesian networks [15].

Mathematically speaking, the Bayesian network compactly represents the joint probability distribution of random variables encoding the domain. The Bayesian network consists of a directed acyclic graph expressing all the necessary dependencies among the random variables, the so-called qualitative part of the model, and the model's local probability distributions parameters, specifying the probabilistic relationships among the random variables, the so-called quantitative part of the model. Each random variable must be conditionally independent of its non-descendants given its parents in the oriented graph.

In the simple case of the "and" relationship among all the parent nodes the individual conditional probability entries can be calculated as

$$\Pr(X_i | parent(X_i)) = \prod_{parent(X_i)} \Pr(parent(X_i))$$

In the simple case of the "or" relationship among all the parent nodes the individual conditional probability entries can be calculated as

$$\Pr(X_i|parent(X_i)) = 1 - \prod_{parent(X_i)} (1 - \Pr(parent(X_i)))$$

The joint probability distribution in factorized form can be represented as follows

$$\Pr(X_1, \ldots, X_n) = \prod_{i=1}^{n} (X_i|parents(X_i))$$

A deeper foundation of probability theory and mathematical statistics is needed to fully understand and apply this modeling approach.

Bayesian networks as a probabilistic framework naturally handles uncertainty in the form of individual health-related factors' description, description of probabilistic relationships between factors, and even the network structure itself at the input. In particular, it provides correctly calculated probabilistic outputs.

The model's graphical nature brings intuitiveness and good understanding by the domain experts, i.e., medical doctors. Nodes of the graph, random variables of interest, represent the health-related factors, for instance, the factors influencing the possible decision. Directed links in the graph can represent statistical as well as causal dependencies among the random variables (Fig. 11.1).

Its probabilistic inference capabilities are critical features for clinical decision support making. The probabilistic inference results can be, for instance, in the form of the so-called posterior distribution of the random variables representing the desired patient's health factor conditioned on the other health-related factors observed in the patient.

There are two basic alternatives for how to construct the Bayesian network. The model can be constructed manually, based on precise knowledge of the medical domain of interest. An irreplaceable role is then played by medical experts in the medical domain and mathematical experts in the field of probabilistic network models.

Bayesian networks can also be learned from the available datasets. This way, we can learn the structure and the parameters of the network without the explicit medical expert knowledge. In this case, a number of prerequisites posed on the data must be met. It is assumed that individual cases comprising the learning dataset are independent, do not influence each other in any way, etc. The critical issue in the learning process is the identification of local conditional dependencies and local conditional independencies among random variables representing the desired factors in the model. There are many approaches how to learn the Bayesian networks currently implemented. Available software packages make it possible to significantly streamline this demanding activity .

The learning process of Bayesian network consists of two phases, learning the graphical structure of the network and learning the parameters of the network, i.e.,

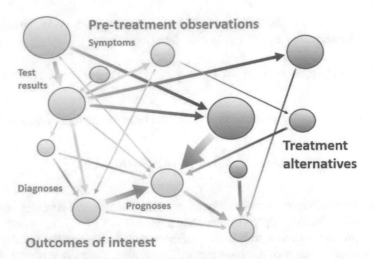

Fig. 11.1 Bayesian networks can serve as a multidimensional decision support framework representing the real-world medical domain. To improve intuitiveness, the nodes can be drawn with a size proportional to their computed importance. The thickness of an arrow can represent the importance of the connection between the nodes

learning the conditional probability distribution associated with each network's node. The prior medical domain knowledge can be incorporated into both learning phases. Due to the compact mathematical representation of the issue, the number of parameters to be learned is significantly lower compared to the number of parameters necessary in the case of the standard joint probability distribution definition.

Due to massive digitization in healthcare, there is an enormous potential of databases comprising the prior domain knowledge, i.e., millions of medical image examinations, historical medical records, known outcomes of prescribed treatment, etc. Past patient records, often from different healthcare institutions where the patient has been treated, provide essential input data for model creation and evaluation. The selection of appropriate cases from the available datasets must be in line with the purpose for which the model is to serve. All the key factors that are supposed to form the final model, including the desired values, must be identifiable in the selection.

There is and still will be a lot of questions to be answered concerning the optimal complexity of the Bayesian network, i.e., the complexity of relationships among the variables representing the factors of the given medical domain under study. In general, the construction of the Bayesian network is an iterative process refining the model step by step, i.e., when additional knowledge or data is coming. It should be noted that in specific situations also some more straightforward modelling techniques need to be considered.

Bayesian networks belong to the family of so-called nonparametric models. As such, the learning process identifying the probabilistic relationships among the random variables representing the desired health-related factors needs a sufficient amount of data. Employing the Bayesian network modelling approach in cases of rare diseases, i.e., having a dataset of limited size, is a challenge. In this situation, the human expert knowledge plays a critical role in the Bayesian network learning process. As such, it can be supplemented by information from relevant complementary sources, literature, and, where possible, combined with more feasible alternative, learning from available data.

Having the Bayesian network of a given medical domain properly established, we can make inferences. The significant advantage of Bayesians networks in the medical domain is their ability to make inferences even with incomplete/missing data, which can be of great importance when modelling the problem areas of rare diseases.

The basic principle of the network is its ability to recalculate the posterior distribution of unobserved desired variables using the prior distribution and having the distribution of observed variables. The patient diagnosis in the Bayesian networks terminology can be informally described as such a value assignment to a subset of variables representing the disease that maximizes their posterior conditional probability distribution given the evidenced value assignment of variables representing the symptoms and other available manifestation of the disease observed in the patient. In other words, a diagnosis of the patient, including its probability, can be interpreted as the most probable assignment of the concerned variables within the model as defined above.

Similarly, the patient treatment prognoses in the Bayesian networks terminology can be informally described as the posterior conditional probability distribution of a set of variables representing the desired aspects of the patient health condition after treatment given the value assignment of variables representing the selected treatment and value assignment of variables representing the symptoms and other available manifestation of the disease observed in the patient.

The above tasks can be extremely computationally intensive, so until recently it was difficult to apply them in practice.

11.6 Conclusion

To summarize, the power of the Bayesian networks, together with the possibilities of global sharing of necessary medical knowledge, represents a promising approach of extracting new, often hidden, knowledge about the given medical domain and thus opens up new ways of achieving the delivery of personalized medicine.

Firstly, the Bayesian networks' ability to capture the existing empirical evidence of a given domain or the disease under study in its complexity and their unique inference mechanisms overcomes the limitations of still prevailing traditional approaches of mathematical statistics. The Bayesian networks allow us to calculate the distribution of the target variable representing the desired health information based on evidenced data of the patient. The target variable can provide personalized predictions, supportive information for personalized decision making, etc. In order to quantify and then compare different treatment strategies, the Bayesian networks' formalism must be extended to include necessary mechanisms from the decision theory.

Secondly, the global cooperation and collection of relevant data for model building eliminate possible misleading results caused by inherent bias when the data comes from only one healthcare institution. The resulting model can be, for instance, validated against the data randomly selected from the databases of different institutions. Incorporating regional and international data into the learning process can further improve the prior distribution and, consequently, the quality of personalized decision support.

This approach benefits from the use of huge international databases, reciprocally then enables the international transmission of complex knowledge about particular medical domains or understanding of diseases under study, including the delivery of meaningful personalized predictions and decision-making support.

One of the main challenges of formal mathematical representation of specific medical domain knowledge is capturing the causality among the factors relevant to diagnostic and prognostic processes. Bayesian networks based on casual relations between the health-related factors enable so-called causal reasoning, i.e., answering causal questions addressing clinical issues. The Bayesian network equipped with a suitable user interface can, for instance, serve as an expert system interactively queried by the diagnosing physician.

The real challenge is discovering something quite new, somehow shift the domain knowledge. In the Bayesian networks terminology, it is called the discovery of latent variables. The Bayesian network is an effective modelling tool for this kind of reasoning. It naturally eliminates the shortcomings of traditional methods of mathematical statistics. Establishing patient diagnosis and treatment prognoses are the critical issues in personalized decision support. Mathematical modelling is beginning to play an irreplaceable role here.

Acknowledgements 1 This publication is based upon work from COST Action "**European Network for cost containment and improved quality of health care-CostCares**" (CA15222), supported by COST (European Cooperation in Science and Technology)

COST (European Cooperation in Science and Technology) is a funding agency for research and innovation networks. Our Actions help connect research initiatives across Europe and enable scientists to grow their ideas by sharing them with their peers. This boosts their research, career and innovation.

https://www.cost.eu

Acknowledgements 2 This work was supported by the "Secure Sharingof Health Care Data During Pandemics—Covid19 (CovidCareShare)".(No.MUNI/33/DP11/2020).

References

1. Sucar, L.E.: Probabilistic Graphical Models—Principals and Applications. Springer, London (2015)
2. Javorník, M., Slavíček, K., Dostál, O.: Sharing of knowledge and resources in regional medical imaging. In: 2nd International Conference on Biomedical Engineering and Assistive Technologies, pp. 304–308. Dr. B. R. Ambedkar National Institute of Technology Jalandhar, Jalandhar, India (2012)
3. Slavíček, K., Javorník, M., Dostál, O.: Extension of the shared regional PACS Center MeDiMed to smaller healthcare institutions. In: Selected Topics on Computed Tomography. IntechOpen, Reunion Island (2012)
4. Word Health Organisation: International Statistical Classification of Diseases and Related Health Problems (ICD) (2019)
5. IHTSDO Systematized Nomenclature of Medicine (SNOMED) Standard
6. NEMA: Digital Imaging and Communications in Medicine (DICOM) Standard (2020)
7. NEMA: DICOM standard's structured report (2013)
8. United Nations: International Covenant on Civil and Political Rights (1976)
9. Council of Europe: Convention for the Protection of Human Rights and Fundamental Freedoms (1950)
10. European Court of Human Rights: Case of I v. Finland (Application no. 20511/03) (2008)
11. Council of Europe: Convention for the Protection of Individuals with regard to Automatic Processing of Personal Data (1981)
12. European Union: Regulation 2016/679/EU on the protection of natural persons with regard to the processing of personal data and on the free movement of such data, and repealing Directive 95/46/EC (General Data Protection Regulation) (2016)
13. European Court of Justice: Case of Patrick Breyer v Bundesrepublik Deutschland, C-582/14 (2016)
14. Paez, M., Tobitsch, K.: The industrial internet of things: risks, liabilities, and emerging legal issues. N. Y. Law School Law Rev. **62**, 217–247 (2017)
15. Fenton, N., Neil, M.: Risk Assessment and Decision Analysis with Bayesian Networks, 1st edn. CRC Press (2013)

Chapter 12
Knowledge-Based UML Dynamic Models Generation from Enterprise Model in Hospital Information Management Process Example

Ilona Veitaite and Audrius Lopata

Abstract The main purpose of this paper is to present knowledge-based Enterprise model (EM) sufficiency as data repository for Unified Modelling Language (UML) models generation. UML models are one of the most usable modelling languages in system lifecycle design stage, despite the problem domain of the system. UML models can be generated from Enterprise Model by using particular transformation algorithms presented in previous researches. Generation process from Enterprise model is represented by certain Hospital Information Management process example. Generated UML dynamic Use Case, Activity, Sequence and State models of different perspectives of Hospital Information Management process prove sufficiency of stored information in Enterprise model.

Keywords UML · Enterprise model · Transformation algorithm · Knowledge-based IS engineering · Hospital IS management

12.1 Introduction

Despite the progress of all information technologies, information system (IS) engineering process still challenges professionals of this field: analysts, designers, researchers and etc. Enterprise modelling makes giant impact to successful information system design process. There are many Enterprise models and Enterprise modelling methodologies, which are applied in different ways and various types of models are built based on chosen Enterprise model [1, 2].

I. Veitaite (✉)
Institute of Social Sciences and Applied Informatics, Vilnius University, Muitinės g. 8, 44280 Kaunas, Lithuania
e-mail: ilona.veitaite@knf.vu.lt

A. Lopata
Kaunas University of Technology, Studentų g. 50, 51368 Kaunas, Lithuania
e-mail: audrius.lopata@ktu.lt

Unified Modelling Language is a highly accepted among IS analysts and designers and is commonly used for IS design. It is used as standard notation to represent designed information system from different views, it provides information in both: structural and behavioural perspectives. Correctly created UML models of any problem domain can be the background for code generation and ensure the success of final IS version [1, 3–5].

Enterprise model can be used as the background for UML models. Correct UML models can be created only then, when gathered into Enterprise Model data is verified, validated and have enough quality. Data gathering process should be done under analysts and experts supervision. Enterprise model with verified and validated data of particular problem domain fully serves all necessary data. Using this data UML models can be generated from Enterprise model through transformation algorithms and after generation process these models are main source for further IS development life cycle stage [3, 4, 6–9].

12.2 Structure of Knowledge-Based Enterprise Model

EMM is formally defined EM structure, which consists of a formalized EM in line with the general principles of control theory. EM is the main source of the necessary knowledge of the particular business domain for IS engineering and IS re-engineering processes (Fig. 12.1) [6, 10].

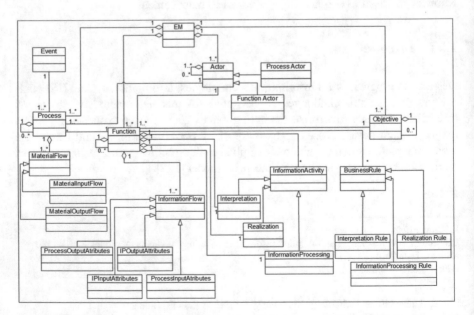

Fig. 12.1 Enterprise meta-model class diagram [6, 9, 10]

EM class model has twenty-three classes. Essential classes are Process, Function and Actor. Class Process, Function, Actor and Objective can have an internal hierarchical structure. These relationships is presented as aggregation relationship. Class Process is linked with the class MaterialFlow as aggregation relationship. Class MaterialFlow is linked with the classes MaterialInputFlow and MaterialOutputFlow as generalization relationship. Class Process is linked with Classes Function, Actor and Event as association relationship. Class Function is linked with classes InformationFlow, InformationActivity, Interpretation, InformationProcessing and Realization as aggregation relationship. These relationships define the internal composition of the Class Function. Class InformationFlow is linked with ProcessOutputAtributes, ProcessInputAtributes, IPInputAttributes and IPOutputAttributs as generalization relationship. Class InformationActivity is linked with Interpretation, InformationProcessing and Realization as generalization relationship. Class Function linked with classes Actor, Objective and BusinessRule as association relationship. Class BusinessRule is linked with Interpretation Rule, Realization Rule, InformationProcessing Rule as generalization relationship. Class Actor is linked with Function Actor and Process Actor as generalization relationship [3, 6, 10, 11].

12.3   Transformation Algorithms of UML Models from Enterprise Model

Each of structural or behavioural UML models can be generated through transformation algorithm and each of models has separate transformation algorithm. These transformation algorithms are presented in previous researches. Main focus of researches is dedicated for generation behavioural or dynamic UML models, because they are more complex and variable [9, 11–13]. To have better understanding of transformation algorithm itself, top level transformation algorithm of UML models generation from EM process is presented in the figure (Fig. 12.2) and described step by step [9, 11–13].

- Step 1: Particular UML model for generation from EM process is identified and selected.
- Step 2: If the particular UML model for generation from EM process is selected then algorithm process is continued, else the particular UML model for generation from EM process must be selected.
- Step 3: First element from EM is selected for UML model, identified previously, generation process.
- Step 4: If the selected EM element is initial UML model element, then initial element is generated, else the other EM element must be selected (the selected element must be initial element).
- Step 5: The element related to the initial element is selected from Enterprise model.

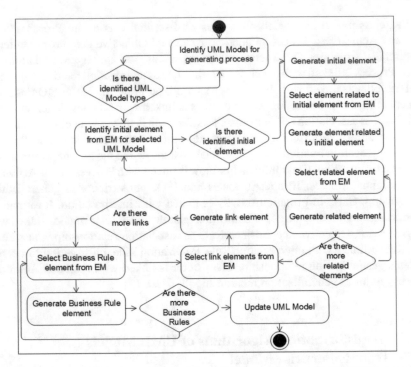

Fig. 12.2 The top level transformation algorithm of UML models generation from EM process [9, 11–13]

- Step 6: The element related to the initial element is generated as UML model element.
- Step 7: The element related to the previous element is selected from Enterprise model.
- Step 8: The element related to the previous element is generated as UML model element.
- Step 9: If there are more related elements, then they are selected from EM and generated as UML model elements one by one, else the link element is selected from Enterprise model.
- Step 10: The link element is generated as UML model element.
- Step 11: If there are more links, then they are selected from EM and generated as UML model elements one by one, else the Business Rule element is selected from Enterprise model.
- Step 12: The Business Rule element is generated as UML model element.
- Step 13: If there are more Business Rules, then they are selected from EM and generated as UML model elements one by one, else the generated UML model is updated with all elements, links and constraints.
- Step 14: Generation process is finished.

Table 12.1 Description of knowledge stored in Enterprise model

Enterprise model element	Description
Actor	In actor element can be stored information related with process or function executor. Actor element is responsible of information related with the process or function participant, it can be person, group of persons, subject such as an IS, subsystem, module and etc
Process, function	In process or function elements can be stored all information related with any user, entity, object, subject and its behaviour. Process or function element is responsible of information related with any operation, activity, status change, movement which is implemented by any actor, entity, participant and etc
Information flow	In Information Flow element can be stored diverse information flow types, such as Information input and output attributes or/and process input and output attributes. Information Flow element is responsible of information related with each element input and output attributes, details which make impact on other elements, their state or status
Business rule	In Business Rule element can be stored different rules such as interpretation, realization or/and information processing. Business rule element is responsible of information about how different elements in IS design phase are related; what restrictions and restraints are applied to these elements

Table 12.1 presents part of Enterprise model elements and their descriptions in order to describe elements, which are necessary in this particular research.

12.4   Generated UML Models of Hospital Information Management Process Example

The Hospital intended to manage outside patients is the object of presented example. In this institution a doctor is only associated with one specialized hospital department (cardiology, pediatrics, etc.) at a time. Each doctor has a visiting time and day in a week.

At reception the patient data is entered and the necessary fees are also taken. The patient is tracked on the basis of the ID number which is generated automatically.

Usually a patient can visit the doctors in two possible ways: directly selecting a doctor or by getting admitted to the hospital.

A doctor can prescribe tests based on the patient's described condition. The patient visits the laboratory to get done the tests prescribed by the doctor. The reports of the tests are given to the patient. The payments related to the tests are done at the reception. According the reports, the doctor prescribes the patient medicines or further tests, if they are needed or is asked to get admitted in hospital.

If available a patient is admitted into a ward of a particular department as per the doctor's prescription. The number of available wards is limited and if there is no free ward the admission of the patient is rescheduled.

Also in case of the prescription of the doctor the patient is operated on a scheduled date and time as decided by the doctor who is responsible for the operation.

After the finishing of the treatment a patient may get discharged on an advice of his doctor and upon the full payment of all due charges at the reception. On payment of complete dues the reception generates a discharge card for the patient.

All data of particular problem domain, in this case, Hospital Information Management data is stored in Enterprise Model described previously. Stored information in Enterprise model is already verified and validated by expert and analyst, so it is ready to use for UML model generation.

12.4.1 UML Use Case Model of Hospital Information Management Process Example

A UML Use Case model is the initial form to identify and present system requirements for a new IS underdeveloped. Use cases identify the expected behaviour—what should be done, and not the exact method of how it should be done. Main advantage of use case modelling is that it assists to design a system from the end user's view. It is a powerful technique for communicating system behaviour in the user's conditions by specifying all externally visible system behaviour [3, 7, 8].

Table 12.2 presents UML Use Case model elements generated from Enterprise model of Hospital Information Management process example. In Enterprise Model all information related with actors, their functions and relationships between these functions is stored. There are four actors: Patient, Doctor, Receptionist and Laboratory Assistant; Receptionist is related with five use cases; Laboratory Assistant—with one use case; Doctor is related with three uses cases and Patient is related with seven use cases. Four use cases includes some additional use cases, six relationships in total. These elements and their relationships are presented in the next figure.

Figure 12.3 presents UML Use Case model of Hospital Information Management process example generated step by step from Enterprise Model through UML Use Case transformation algorithm.

12.4.2 UML Activity Models of Hospital Information Management Process Example

UML Activity model describes how activities are coordinated, activities dependence from the actor or previous activity. It provides a service which can be in various levels of abstraction. Usually, an event needs to be gained by some operations, particularly

Table 12.2 UML use case model elements generated from enterprise model of hospital information management process example [3, 7, 8]

Enterprise model element	UML use case model element	Hospital information management process example	Description
Actor	Actor	Patient	There are four actors, each of them is behavioural classifier which defines a role played in particular example
		Doctor	
		Receptionist	
		Laboratory assistant	
Process, function	Use case	Laboratory visit for the test	There are fourteen use cases, each use case is a type of behavioural classifier that describes a unit of functionality performed by three actors
		Test report generation	
		Payment for the test at reception	
		Registration for treatment	
		ID generation	
		Fee payment	
		Admission to ward	
		Discharging	
		Account settlement	
		Discharging card generation	
		Test prescription	
		Test report analysis	
		Prescription for medicines	
		Operation performing	
Business rule	Include	Six include elements	There are six include elements, each include is a directed relationship between two use cases which is used to demonstrate that behaviour of the included use case is inserted into the behaviour of the including use case

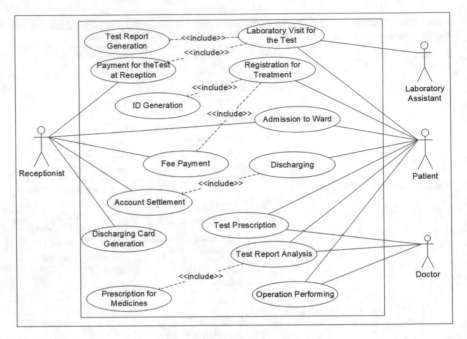

Fig. 12.3 UML use case model of hospital information management process example

where the operation is intended to gain a number of different things that require coordination, or how the events in a single use case relate to one another, especially, use cases where activities may overlap and require coordination [3, 7, 8].

According previously described UML Use Case model there is possible to identify at least five different UML Activity models: Patient Registration, Ward Assignation, Medical Tests, Treatment Process and Discharging.

UML Activity Model: Patient Registration

First UML Activity Model generated from EM is Patient Registration, where two participants—actors take part: Patient and Receptionist.

Table 12.3 presents UML Activity model elements generated from Enterprise model of Hospital Information Management process example, Registration part. Actor—first UML Activity model partition Patient starts registration process: visits reception, provides personal data, Actor—second partition Receptionist enters patient's data and provides patient's ID number, last activity Fee Payment is related with first partition, Patient pays the fee and registration process ends.

Figure 12.4 presents UML Activity Model of Hospital Information Management process example, Registration part generated step by step from Enterprise model through UML Activity model transformation algorithm [9].

Table 12.3 UML activity model elements generated from enterprise model of hospital information management process example, registration part [3, 7, 8]

Enterprise model element	UML activity model element	Hospital information management process example	Description
Actor	Partition	Patient	There are two partitions and activities are related with these actors
		Receptionist	
Function, process	Activity	Reception visit	There are five activities directly related with two partitions: patient—three activities, receptionist—two. They represent a parameterized behaviour as coordinated flow of actions
		Personal data provision	
		Data entering into system	
		Patient ID generation	
		fee payment	
Business rules	Control nodes	Initial node	There are two control nodes: one node—initial node in the beginning; final node in the end of the process
		Final node	

Fig. 12.4 UML activity model of hospital information management process example. registration

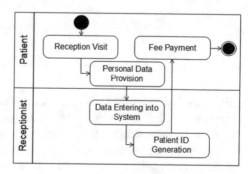

UML Activity Model: Ward Assignation

Second UML Activity Model generated from EM is Ward Assignation, where two participants—actors take part: Patient and Receptionist.

Table 12.4 presents UML Activity model elements generated from Enterprise model of Hospital Information Management process example, Ward Assignation

Table 12.4 UML activity model elements generated from enterprise model of hospital information management process example, ward assignation part [3, 7, 8]

Enterprise model element	UML activity model element	Hospital information management process example	Description
Actor	Partition	Patient	There are two partitions and activities are related with these actors
		Receptionist	
Function, process	Activity	Ward availability check	There are four activities directly related with two partitions: patient—one activity, receptionist—three. They represent a parameterized behaviour as coordinated flow of actions
		Provision of new dates	
		Ward assignment	
		New dates inquiry	
Information flow	Object flow edge	Ward assignation details to patient	There are two object flow edges which are activity edges used to show data flow between activities
		Ward information update to reception	
Business rules	Control nodes	Initial node	There are five control nodes: one node—initial node in the beginning; decision node—for ward availability check; join and fork nodes—to relate object flow edges, final node in the end of the process
		Decision node	
		Join node	
		Fork node	
		Final node	

part. Actor—first UML Activity model partition Receptionist starts Ward Assignation process: Checks ward availability, assigns it, or inquires for new dates, because there are no free wards, Actor—second partition Patient provides new date for ward assignation, last activities are related with first partition, Receptionist prepares information for patient and updates information in Reception and process ends.

Figure 12.5 presents UML Activity Model of Hospital Information Management process example, Ward Assignation part generated step by step from Enterprise model through UML Activity model transformation algorithm [9].

UML Activity Model: Medical Tests

Third UML Activity Model generated from EM is Medical Tests, where three participants—actors take part: Patient, Laboratory Assistant and Receptionist.

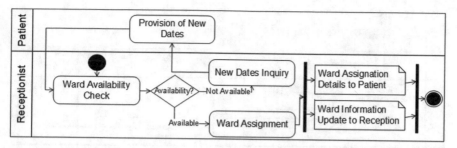

Fig. 12.5 UML activity model of hospital information management process example. Ward assignation

Table 12.5 UML activity model elements generated from enterprise model of hospital information management process example, medical tests part [3, 7, 8]

Enterprise model element	UML activity model element	Hospital information management process example	Description
Actor	Partition	Patient	There are three partitions and activities are related with these actors
		Laboratory assistant	
		receptionist	
Function, process	Activity	Laboratory visit for test	There are ten activities directly related with three partitions: patient—four activities, laboratory assistant—five activities, receptionist—one. They represent a parameterized behaviour as coordinated flow of actions
		Doctor's prescription check	
		Sample inquiry	
		Sample provision	
		Performing of the test	
		Payment order generation	
		Report generation	
		Fee payment	
		Issuing receipt	
		Payment receipt provision	
Business rules	Control nodes	Initial node	There are four control nodes: one node—initial node in the beginning; join and fork nodes—to relate additional activities; final node in the end of the process
		Join node	
		Fork node	
		Final node	

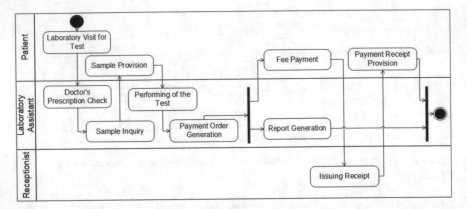

Fig. 12.6 UML activity model of hospital information management process example. Medical tests

Table 12.5 presents UML Activity model elements generated from Enterprise model of Hospital Information Management process example, Medical Tests part. Actor—first UML Activity model partition Patient starts Medical Tests process: visits laboratory and provides sample after inquiry, Actor—second partition Laboratory checks doctor's prescription, inquires for sample, performs test, generates payment order and prepares report for the doctor; Actor—third partition Receptionist confirms payment form the patient and provides receipt; Patient makes payment and after payment confirmation receives receipt and process ends.

Figure 12.6 presents UML Activity Model of Hospital Information Management process example, Medical Tests part generated step by step from Enterprise model through UML Activity model transformation algorithm [9].

UML Activity Model: Treatment Process

Fourth UML Activity Model generated from EM is Medical Tests, where two participants—actors take part: Patient and Doctor.

Table 12.6 presents UML Activity model elements generated from Enterprise model of Hospital Information Management process example, Treatment Process part. Actor—first UML Activity model partition Doctor starts Treatment Process: meets the patient, analyses provided test reports, regarding test results decides to discharge patient or continue treatment process. Doctor decides if there is need to do more tests or not, assigns treatment method medicine or operational intervention, after actor—second partition Patient confirmation, Doctor performs operation and process ends.

Figure 12.7 presents UML Activity Model of Hospital Information Management process example, Treatment Process part generated step by step from Enterprise model through UML Activity model transformation algorithm [9].

Table 12.6 UML activity model elements generated from enterprise model of hospital information management process example, treatment process part [3, 7, 8]

Enterprise model element	UML activity model element	Hospital information management process example	Description
Actor	Partition	Patient	There are two partitions and activities are related with these actors
		Doctor	
Function, process	Activity	Patient visit	There are ten activities directly related with two partitions: patient—two activities, doctor—eight. They represent a parameterized behaviour as coordinated flow of actions
		Test report provision	
		Report analysis	
		Issuing discharge	
		Test requirements check	
		Test prescription	
		Treatment requirement check	
		Operation scheduling	
		Confirmation of operation	
		Performing operation	
Business rules	Control nodes	Initial node	There are five control nodes: one node—initial node in the beginning; three decision nodes—for test report status, for more tests possibility, for treatment type; final node in the end of the process
		Decision nodes	
		Final node	

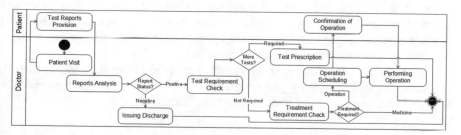

Fig. 12.7 UML activity model of hospital information management process example. Treatment process

Table 12.7 UML activity model elements generated from enterprise model of hospital information management process example, discharging part [3, 7, 8]

Enterprise model element	UML activity model element	Hospital information management process example	Description
Actor	Partition	Patient	There are two partitions and activities are related with these actors
		Receptionist	
Function, process	Activity	Approaching with discharge advice	There are six activities directly related with two partitions: patient—two activities, receptionist—four. They represent a parameterized behaviour as coordinated flow of actions
		Data check	
		Discharge card generation	
		Payment check order	
		Due amount payment	
		Discharge card provision	
Business rules	Control nodes	Initial node	There are two control nodes: one node—initial node in the beginning; decision node—for payment status final node in the end of the process
		Decision node	
		Final node	

UML Activity Model: Discharging

Fifth UML Activity Model generated from EM is Discharging, where two participants—actors take part: Patient and Receptionist.

Table 12.7 presents UML Activity model elements generated from Enterprise model of Hospital Information Management process example, Discharging part. Actor—first UML Activity model partition Patient starts discharging process: approaches with discharge advice from the doctor, Actor—second partition Receptionist checks data, generate discharge card, check payment status, after patient makes payment, provides discharge card and process ends.

Figure 12.8 presents UML Activity Model of Hospital Information Management process example, Discharging part generated step by step from Enterprise model through UML Activity model transformation algorithm [9].

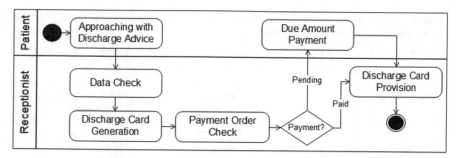

Fig. 12.8 UML activity model of hospital information management process example. Discharging

12.4.3   UML Sequence Models of Hospital Information Management Process Example

UML Sequence model is an interaction model that defines how operations are implemented. This model fixates the interaction between objects in the context of a collaboration. UML Sequence model is time focus and it shows the order of the interaction visually by using the vertical axis of the diagram to deliver time what messages are sent and when [3, 7, 8].

According previously described UML Use Case and UML Activity models there is possible to identify at least three different UML Sequence models: Patient Admission, Tests and Treatment, and Discharging.

UML Sequence Model: Patient Admission

First UML Sequence Model generated from EM is Patient Admission, where four participants—Lifelines take part: Patient, Receptionist, Database and Ward.

Table 12.8 presents UML Sequence model elements generated from Enterprise model of Hospital Information Management process example, Patient Admission part. In Enterprise Model all information related with actors and their collaboration is stored. There are four actors—process participants, which are called Lifelines in UML Sequence model: persons—Patient, Receptionist, subject—Database, object—Ward. Patient registers to the hospital, Receptionist enters gathered data, Patient requests for the ward, Receptionist checks availability and confirms or denies ward availability.

Figure 12.9 presents UML Sequence model of Hospital Information Management process example, Patient Admission part generated step by step from Enterprise model through UML Sequence model transformation algorithm [9].

UML Sequence Model: Tests and Treatment

Second UML Sequence Model generated from EM is Tests and Treatment, where four participants—Lifelines take part: Patient, Doctor, Operation and Test.

Table 12.9 presents UML Sequence model elements generated from Enterprise model of Hospital Information Management process example, Test and Treatment

Table 12.8 UML sequence model elements generated from enterprise model of hospital information management process example, patient admission [3, 7, 8]

Enterprise model element	UML sequence model element	Hospital information management process example	Description
Actor	Lifeline	Patient	There are four actors, in UML Sequence model four Lifelines, which are shown using a symbol that consists of a rectangle forming its "head" followed by a vertical line and these lines represent the lifetime of the actor—participant of the process
		Receptionist	
		Database	
		Ward	
Process, function	Message	Register(data)	There are eleven messages, related with actors and they define a communication between these actors
		Addnew(data)	
		Return	
		Return	
		Wardrequest()	
		Availabilitycheck()	
		Return(status)	
		[not available] return(n/a)	
		[if available] wardupdate(data)	
		Return	
		Return(noward)	
Business rules	Execution specification	Five execution specifications	Each of five executions specification element represents a period in the actor's lifetime

part. In EM all information related with actors-lifelines and their collaboration is
stored. There are four actors—process participants, which are called Lifelines in
UML Sequence model: persons—Patient, Receptionist, objects—Operation, Test.
Doctor performs check-up and prescribes medicine, if necessary prescribes test,
Patient provides samples and gets reports, Doctor reviews reports and prescribes more
medicine or prescribes operation if necessary, prescribes more tests and operates.

Figure 12.10 presents UML Sequence model of Hospital Information Management process example, Test and Treatment part generated step by step from Enterprise
model through UML Sequence model transformation algorithm [9].

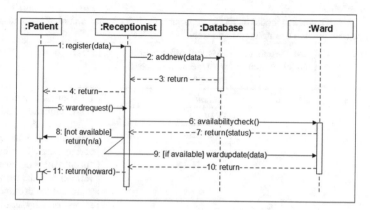

Fig. 12.9 UML sequence model of hospital information management process example, patient admission

UML Sequence Model: Discharging

Third UML Sequence Model generated from EM is Discharging, where five participants—Lifelines take part: Doctor, Patient, Reception, Database and Ward.

Table 12.10 presents UML Sequence model elements generated from Enterprise model of Hospital Information Management process example, Discharging part. In EM all information related with lifelines and their cooperation is stored. There are five actors—process participants, which are called Lifelines in UML Sequence model: persons—Patient, Doctor, subjects—Reception, Database, object—Ward. Doctor provides discharge advice, Patient requests for discharge, Reception checks information related with payments, inquires for payment, Patient makes the payment, Reception updates financial information in Database, provides Receipt, Reception updates discharge information and information related with ward, in the end Reception provides Discharge card.

Figure 12.11 presents UML Sequence model of Hospital Information Management process example, Discharging part generated step by step from Enterprise model through UML Sequence model transformation algorithm [9].

12.4.4   UML State Models of Hospital Information Management Process Example

UML State Model demonstrates the diverse states of an entity. State model can also show how an entity responds to various events by changing from one state to another [3, 7, 8].

According previously described UML models there is possible to identify at least three different UML State models describing states of: Patient, Doctor and Ward.

Table 12.9 UML sequence model elements generated from enterprise model of hospital information management process example, test and treatment [3, 7, 8]

Enterprise model element	UML sequence model element	Hospital information management process example	Description
Actor	Lifeline	Patient	There are four actors, in UML Sequence model four Lifelines, which are shown using a symbol that consists of a rectangle forming its "head" followed by a vertical line and these lines represent the lifetime of the actor—participant of the process
		Doctor	
		Operation	
		test	
Process, function	Message	Performcheckup()	There are thirteen messages, related with actors and they define a communication between these actors
		Return	
		Prescribemedicine()	
		Return	
		Prescribetest()	
		Providesamples(samples)	
		Return(report)	
		Inquirereview(reports)	
		Prescribemedicine()	
		Prescribeoperation()	
		Moretest()	
		Getoperated()	
		Operate()	
Business rules	Execution specification	Five execution specifications	Each of five executions specification element represents a period in the actor's lifetime

UML State Model: Patient

First UML State Model generated from EM describes states of Patient.

Table 12.11 presents UML State model elements generated from Enterprise model of Hospital Information Management process example, Patient part. In Enterprise Model all information related with processes, functions and their states is stored. This model is from Patient's perspective. In the model these elements are presented: initial state which starts the process, first state Patient registered, it's state changes after doctors visit: patient receives treatment, additional doctor's visit, after which doctor advices discharge procedure and patient's state changes again, patient is discharged, process ends with final state.

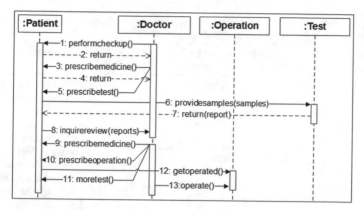

Fig. 12.10 UML sequence model of hospital information management process example, test and treatment

Figure 12.12 presents UML State model of Hospital Information Management process example, Patient part.

UML State Model: Doctor

Second UML State Model generated from EM describes states of Doctor.

Table 12.12 presents UML State model elements generated from Enterprise model of Hospital Information Management process example, Doctor part. In Enterprise Model all information related with processes, functions and their states is stored. This model is from Doctor's perspective. In the model these elements are presented: initial state which starts the process, first state Doctor registered, it's state changes after patient registers to the visit: Doctor prescribes treatment, check's up treatment results, after patient's discharge procedure Doctor's state changes, he is not needed for this particular patient, process ends with final state.

Figure 12.13 presents UML State model of Hospital Information Management process example, Doctor part.

UML State Model: Ward

Third UML State Model generated from EM describes states of Ward.

Table 12.13 presents UML State model elements generated from Enterprise model of Hospital Information Management process example, Ward part. In Enterprise Model all information related with processes, functions and their state is stored. This model is from Ward's perspective. In the model these elements are presented: initial state which starts the process, first state means ward is free, its state changes after request to occupy; after patient is discharged ward state changes again to free.

Figure 12.14 presents UML State model of Hospital Information Management process example, Ward part.

With the help of Hospital Information Management process example result of four UML models: Use Case, Activity, Sequence and State generation from Enterprise Model through transformation algorithms is presented in detailed way, all models

Table 12.10 UML sequence model elements generated from enterprise model of hospital information management process example, discharging [3, 7, 8]

Enterprise model element	UML sequence model element	Hospital information management process example	Description
Actor	Lifeline	Doctor	There are five actors, in UML Sequence model four Lifelines, which are shown using a symbol that consists of a rectangle forming its "head" followed by a vertical line and these lines represent the lifetime of the actor—participant of the process
		Patient	
		Reception	
		Database	
		Ward	
Process, function	Message	Dischargeadvice()	There are fifteen messages, related with actors and they define a communication between these actors
		Requestdiscgarge()	
		Checkdues(patientid)	
		Return(dues)	
		Askpayment(amount)	
		Paydues(amount,patientid)	
		Update(amount,patientid)	
		Return(receipt)	
		Return(receipt)	
		Return	
		Updatedischargedata(patientid)	
		Updateward()	
		Return	
		Return	
		Grantdicharge(dichargecard)	
Business rules	Execution specification	Ten Execution Specifications	Each of ten executions specification element represents a period in the actor's lifetime. Each parallels defines potentially parallel execution of behaviors of the operands of the combined fragment
	Parallel	Two parallels	

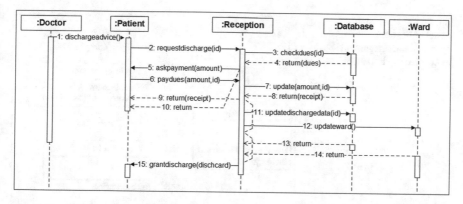

Fig. 12.11 UML sequence model of hospital information management process example, discharging

Table 12.11 UML state model elements generated from enterprise model of hospital information management process example, patient [3, 7, 8]

Enterprise model element	UML state model element	Hospital information management process example	Description
Process, function	Transition	Doctor visit	Transitions from one state to the next respond to the activities, events, what causes the state's change
		Doctor review	
		Issue to discharge	
Information flow	Simple state	Patient registered	Internal activities compartment holds a list of internal actions or state (do) activities (behaviours) that are performed while the element is in the state
		Treatment in progress	
		Discharged	
Business rules	Initial and final states	Initial state	Initial and final states are a special kind of states signifying the beginning and closing processes of defined states
		Final state	

define same example, but from different perspectives. In almost each subsection of the described example there are more than one model of the same type presented: generated UML Use Case model presents all participants (Actors), which are involved in Hospital Information Management process and their functions/processes (Use Cases); generated UML Activity models illustrate different activities from different perspectives (Registration, Ward Assignation, Medical Tests, Treatment process and Discharging) of the same example, and it is not final list of possible models of

Fig. 12.12 UML state
model of hospital
information management
process example, patient

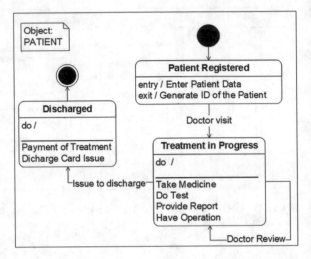

Table 12.12 UML state model elements generated from enterprise model of hospital information management process example, doctor [3, 7, 8]

Enterprise model element	UML state model element	Hospital information management process example	Description
Process, function	Transition	Patient Registered	Transitions from one state to the next respond to the activities, events, what causes the state's change
		Patient (re)Checkup	
		Planned leave of the Doctor	
Information flow	Simple state	Doctor registered	Internal activities compartment holds a list of internal actions or state (do) activities (behaviours) that are performed while the element is in the state
		Appointing treatment	
		Doctor inactive	
Business rules	Initial and final states	Initial state	Initial and final states are a special kind of states signifying the beginning and closing processes of defined states
		Final state	

the same type; generated UML Sequence model also define sequence processes and functions sequences from different perspectives (Patient Admission, Tests and Treatment, Discharging) of the same example, and it is also not final list of possible UML Sequence models; generated UML State model describe different states from the perspectives of objects (Patient, Doctor and Ward), and states of more objects of the same example can be generated.

Fig. 12.13 UML state model of hospital information management process example, patient

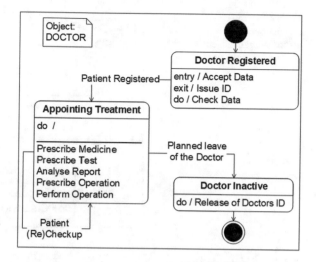

Table 12.13 UML state model elements generated from enterprise model of hospital information management process example, Ward [3, 7, 8]

Enterprise model element	UML state model element	Hospital information management process example	Description
Process, function	Transition	Request to occupy	Transitions from one state to the next respond to the activities, events, what causes the state's change
		Patient discharged	
Information flow	Simple state	Free	Internal activities compartment holds a list of internal actions or state (do) activities (behaviours) that are performed while the element is in the state
		Occupied	
Business rules	Initial and final states	Initial state	Initial state is a special kind of states signifying the beginning process of defined states

Provided example of Hospital Information Management process shows and confirms, that it is not the final amount of UML models, which can be generated from EM, there are more different perspectives for UML models generation of the same example. As stated previously, knowledge-based Enterprise model which stores verified and validated data of a specific problem domain is enough data storage for generation various UML models.

Fig. 12.14 UML state
model of hospital
information management
process example, ward

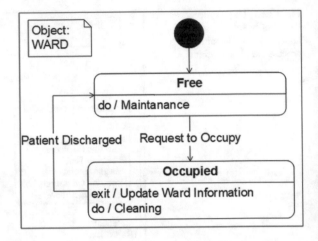

12.5 Conclusions

The first part of the paper presents the structure of knowledge-based Enterprise model, by defining all its components and their relations.

The second part deals with the presentation of UML models generation form Enterprise model top level transformation algorithm, which is defined structurally step by step. In this section also part of Enterprise model elements, necessary for particular example of the research, are presented and their descriptions are also provided.

The third part presents Hospital Information Management process example, which data is stored in knowledge-based Enterprise model and used for UML models generation. There are defined four types of UML dynamic models in details, which represent data of chosen example in different perspectives. Each presented UML model is generated through certain transformation algorithms introduced in previous researches.

Each subsection describes different type of UML dynamic model and presents particular type of UML models variations: same UML model type, but different data used for different perspectives. All these UML models are generated based the knowledge stored in Enterprise model.

The defined Hospital Information Management process example shows that verified and validated knowledge stored in Enterprise model is sufficient for UML models generation process; that stored in Enterprise model elements are enough to transfer all UML models elements, despite the perspective of certain UML model. Using UML models generated from Enterprise Model full IS development life cycle design stage can be implemented as knowledge-based process.

Acknowledgements This publication is based upon work from COST Action "**European Network for cost containment and improved quality of health care-CostCares**" (CA15222), supported by COST (European Cooperation in Science and Technology)

COST (European Cooperation in Science and Technology) is a funding agency for research and innovation networks. Our Actions help connect research initiatives across Europe and enable scientists to grow their ideas by sharing them with their peers. This boosts their research, career and innovation.

https://www.cost.eu

References

1. Dunkel, J., Bruns, R.: Model-driven architecture for mobile applications. In: Proceedings of the 10th Inter-national Conference on Business Information Systems (BIS), Vol. 4439/2007, pp. 464–477 (2007)
2. Eichelberger, H., Eldogan, Y., Schmid, K.A.: Comprehensive Analysis of UML Tools, their Capabilities and Compliance. Software Systems Engineering. Universität Hildesheim. versio 2.0 (2011)
3. Jacobson, I., Rumbaugh, J., Booch, G.: Unified Modeling Language User Guide, 2nd edn. Addison-Wesley Professional. ISBN: 0321267974 (2005)
4. Jenney, J.: Modern methods of systems engineering: with an introduction to pattern and model based methods. ISBN-13:978-1463777357 (2010)
5. Sajja, P.S., Akerkar, R.: Knowledge-Based Systems for Development. Advanced Knowledge Based Systems: Model, Applications & Research, Vol. 1 (2010)
6. Gudas, S.: Architecture of Knowledge-Based Enterprise Management Systems: a Control View. In: Proceedings of the 13th world multiconference on systemics, cybernetics and informatics (WMSCI2009),) July 10–13, Orlando, Florida, USA, Vol. III, pp. 161–266. ISBN -10: 1-9934272-61-2 (Volume III). ISBN -13: 978-1-9934272-61-9 (2009)
7. OMG UML (2019) Unified Modeling Language version 2.5.1. Unified Modelling. https://www.omg.org/spec/UML/About-UML/
8. UML Diagrams: UML diagrams characteristic (2012). http://www.uml-diagrams.org
9. Veitaitė, I., Lopata, A.: Transformation algorithms of knowledge based UML dynamic models generation. Business information systems workshops BIS 2017, Poznan, Poland, 28–30 June. Abramowicz, W, (ed.), Lecture Notes in Business Information Processing, Vol. 303. Springer International Publishing, Cham (2017)
10. Gudas, S.: Informacijos sistemų inžinerijos teorijos pagrindai/Fundamentals of Information Systems Engineering Theory. (Lithuanian) Vilnius University. ISBN 978-609-459-075-7 (2012)
11. Veitaitė, I., Lopata, A.: Problem domain knowledge driven generation of UML models. In: Damaševičius, R., Vasiljevienė, G. (eds.), Information And Software Technologies: 24th International Conference, ICIST 2018, Vilnius, Lithuania, October 4 6, 2018: Proceedings. Springer, Cham (2018)
12. Veitaite, I., Lopata, A.: Knowledge-Based Transformation Algorithms of UML Dynamic Models Generation from Enterprise Model. In: Dzemyda, G., Bernatavičienė, J., Kacprzyk, J. (eds.), Data Science: New Issues, Challenges and Applications. Studies in Computational Intelligence, Vol. 869. Springer, Cham (2020)
13. Veitaite, I., Lopata, A.: Knowledge-Based Generation of the UML Dynamic Models from the Enterprise Model Illustrated by the Ticket Buying Process Example. In: Lopata, A., Butkienė, R., Gudonienė, D., Sukackė, V. (eds.), Information and Software Technologies. ICIST 2020. Communications in Computer and Information Science, Vol. 1283. Springer, Cham (2020). https://doi.org/10.1007/978-3-030-59506-7_3

Printed in the United States
by Baker & Taylor Publisher Services